FIREFLY

# DEEP SKY
## OBSERVER'S GUIDE

### NEIL BONE

FIREFLY BOOKS

# A FIREFLY BOOK

Published by Firefly Books Ltd. 2005

First printing

Publisher Cataloging-in-Publication Data

Bone, Neil.
    Deep sky observer's guide / Neil Bone. – 1st ed.
[224] p. : ill., photos. (some col.), charts, maps ;   cm.
Includes bibliographical references and index.
Summary: Introduction to deep sky observing using binoculars
and small telescopes.
ISBN 1-55407-024-4 (pbk.)
1. Astronomy--Observers' manuals. 2. Telescopes. I. Title.
523  dc22    QB64.B664 2005

Library and Archives Canada Cataloguing in Publication

Bone, Neil, 1959-
        Deep sky observer's guide / Neil Bone.

Includes bibliographical references and index.
ISBN 1-55407-024-4

        1. Astronomy--Observers' manuals. I. Title.

QB64.B65 2005          522          C2004-904692-6

Published in the United States in 2005 by
Firefly Books (U.S.) Inc.
P.O. Box 1338, Ellicott Station
Buffalo, New York 14205

Published in Canada in 2005 by
Firefly Books Ltd.
66 Leek Crescent
Richmond Hill, Ontario L4B 1H1

Published in Great Britain in 2004 by Philip's,
a division of Octopus Publishing Group Ltd,
2–4 Heron Quays, London E14 4JP

Printed in China

*Front cover:* Rho Ophiuchi
Cloud (Stephen Pitt). In the
foreground is a reflector on
an equatorial mount.

*Back cover:* (above) Spiral
galaxy M74 (Todd
Boroson/NOAO/AURA/NSF).

(below) Finder chart for Virgo
galaxies (John Cox/Richard
Monkhouse/Philip's).

*Title Page:* North America
Nebula (Ian King).

# CONTENTS

# I · INTRODUCING THE DEEP SKY

A stronomy is surely the most visual of the sciences, and this is largely why it has so captured the public imagination. Television news reports and the daily papers often carry the latest images from space exploration. Pictures of Mars, for example, were prominent on the front pages in August 2003, during the Red Planet's closest approach for 60,000 years, followed by extensive media coverage of the Spirit and Opportunity rover landings in early 2004. Of the great many images from the orbiting Hubble Space Telescope (HST) to have entered the public domain, perhaps none enjoys more lingering iconic status than that of the Eagle Nebula, released in 1995. The HST's clear view from above the Earth's distorting atmosphere revealed the nebula in unprecedented detail. The nebula's "Pillars of Creation" – long fingers of dark dust tipped by newly formed stars – have become as familiar as the Voyager images of the outer planets.

The Eagle Nebula is found in the constellation of Serpens, just north of the richest parts of the Milky Way in Sagittarius and Scorpius. While it can never be seen in such fine detail through a typical amateur astronomer's telescope as it was by the HST's sophisticated instrumentation, the Eagle Nebula (also known as M16) is a familiar sight for many whose passion is observing the deep sky.

◄ The 1995 Hubble Space Telescope portrait of the Eagle Nebula in Serpens shows the dark "Pillars of Creation" associated with star formation in this distant cloud of gas and dust. It has become one of the most familiar astronomical images of our time.

Quite when and where the term "deep sky" came into widespread use is unclear. It has certainly been used at least since the late 1960s as an umbrella description for objects far beyond the bounds of the Solar System. As an activity for amateur astronomers, deep sky observing has often been regarded as more of a pastime than an exercise in obtaining potentially useful scientific data, such as brightness estimates for variable stars, meteor counts, or details of changes in Jupiter's cloud belts. It does remain a largely recreational activity: amateurs' telescopic observations are never going to add new objects to the existing catalogs, for example, though it

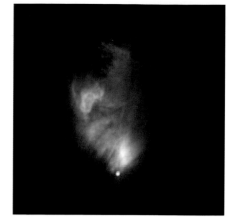

▲ American amateur astronomer Jay McNeil found this patch of nebulosity near M78 in Orion on images he took early in 2004. Such new discoveries in the deep sky are rare.

was notably an amateur observer who in early 2004 first detected a new bright patch, which soon became known as McNeil's Nebula, in the reflection nebula M78 in Orion (the "M" number is the nebula's designation in the best-known listing, or catalog, of deep sky objects). There is, however, a large and growing band of observers who take their deep sky studies a little more seriously, using faint, hard-to-find objects as a test of telescope performance, visual acuity and ability to navigate around the night sky.

Skills acquired by the experienced deep sky observer are applicable to other fields of astronomy. Perhaps most obviously, hunting out faint comets – in appearance, "mobile nebulae" – demands exactly the same skills as finding faint deep sky objects.

## Discoveries in the CCD age

While extensive sky surveys have long since swept up all the amateur-detectable deep sky objects, discoveries remain to be made. Some dedicated deep sky enthusiasts have established observing programs to search for supernova explosions in distant galaxies. Newly discovered supernovae offer professional astronomers valuable insights into the scale and evolution of the Universe, and amateurs carrying out semiautomated searches can help by providing early alerts. Some – most notably the Rev. Robert Evans in Australia – have made important

▲ *In recent years, patrols by amateur astronomers using CCD cameras have led to a great many discoveries of supernovae in distant galaxies. Discovered by UK observer Mark Armstrong, SN 2001ef appears here as a previously unseen star close to the nucleus of the faint galaxy IC 381.*

supernova discoveries using only visual telescopic searches. Such searches demand a familiarity with the normal appearance of many galaxies if the presence of any "new" star is to be immediately apparent to the observer. The most prolific amateur supernova discoverers have in recent years used automated telescopes which are capable of rapidly locating upward of 30 galaxies an hour and are equipped with CCD (charge coupled device) cameras that take "patrol" images. The images are later compared with a reference image to check for the presence of potential interloper stars. Mark Armstrong and Tom Boles in the UK have between them discovered well over a hundred supernovae in this way, while searches in the United States by Michael Schwartz, Tim Puckett and others associated with teams such as LOTOSS, guided by professional astronomers at institutions such as the Lowell Observatory, have similarly proved fruitful.

CCD cameras have revolutionized deep sky imaging. Their high sensitivity to light makes them much more efficient than photographic film for recording faint objects, and with many good post-observation image processing programs now available for personal computers, digital imaging is already beginning to replace film as the medium of choice for imaging the deep sky. Whereas traditional deep sky photography often demanded expensive (and, in the case of gas hypersensitization procedures, sometimes dangerous!) pretreatment of the film and long exposure times, CCDs can record more at the touch of a button and in a matter of seconds.

One downside to CCD imaging is the expense of the hardware. When CCD equipment for amateur astronomers first appeared on the market, imaging chips were small and costly. Size is gradually increasing and prices have come down, but setting up to do CCD imaging still demands a sizable outlay: a laptop computer, at the very least, is needed to run the imaging software and store raw data, on top of the cost of the CCD camera itself. In this book I concentrate on visual observing and recording by simple methods. I am quite sure, though, that after an initial visual exploration of the deep sky, some readers will feel encouraged to take their interest further, into the realm of photographic and CCD imaging.

## The Visual Deep Sky

At the other end of the observational spectrum, there is absolutely no reason why locating deep sky objects shouldn't be simply a pleasant recreation! In common with many others, I enjoy those occasional crisp, clear nights with a dark sky and a couple of spare hours to spend searching out a nebula, star cluster or galaxy which I have never looked at through a telescope – or revisiting "old friends" which I may not have seen for a few years. Amateur astronomy is pursued in most cases for the enjoyment and satisfaction it brings, and the observation of interesting deep sky objects should certainly provide both.

Many detailed handbooks and guides have appeared over the years, listing the huge numbers of faint deep sky objects that can be glimpsed through large telescopes (sometimes described crudely as "light buckets"). My aim here, though, is to provide an introduction for observers with smaller instruments. Most of the targets I describe are reasonably bright and easy to find, but I also throw in a few more challenging objects to add to the thrill of the chase. All the objects included in this book should be within the reach of a small 80 to 100 mm aperture refractor or a 150 mm reflector, typical portable instruments owned by the beginning or moderately experienced observer.

Although I have been observing the sky for more than 30 years, I still find it convenient to do much of my occasional deep sky observing with a wide-field 80 mm aperture refractor – essentially a "spotting scope." It can be set up for observing and packed away afterward in a matter of minutes – an important consideration at the end of a long working day, or when there is only a little time available for viewing.

Binoculars can give good views of brighter objects and are useful in a number of other areas of practical amateur astronomy – variable star observing, for example. Even if the observer has access to a small telescope, binoculars are useful for obtaining a low-power view of the field and a first impression of the target. Even quite small instruments, such as 10 × 50 binoculars, will provide observers with their own, first-hand views of the more prominent deep sky objects such as the Eagle Nebula. While they may not reveal the awesome detail of HST images, these instruments offer the satisfaction of seeing things for oneself.

## Meet the Family

By definition, deep sky objects are those that lie beyond the Solar System. Some of the objects covered in this book are very remote indeed. Astronomers find it convenient to deal with distances within the Solar System in terms of the astronomical unit (AU), which is equivalent to the mean orbital distance of the Earth from the Sun

(150,000,000 km). Jupiter, the largest of the planets, is a little over 5 AU from the Sun, while Neptune, the most distant of the Sun's sizable planets, orbits at around 30 AU.

Such distances are minuscule, however, when compared with the distances that separate the stars. For interstellar measurements, astronomers consider the time that light takes to travel from an object. The laws of physics dictate that nothing in the Universe can travel faster than the speed of light, which is 300,000 km/s. Among the interesting consequences of this physical barrier is the fact that we see distant objects not as they are now, but as they were when the light that we detect left them – a concept known as *lookback time*.

In terms of lookback time, distances in the Solar System are small. The average Earth–Moon distance (384,000 km) equates to 1.3 light seconds. The Sun is 8.3 light minutes away. Rather more substantial is the 33-minute light-time to Jupiter when the planet is at its closest, and once we get to Neptune, the distance is more like 4 light hours. By comparison, the distances to nearby stars have to be reckoned in light *years*. The closest stellar system is the southern double star Alpha Centauri and its dim but famous companion Proxima, 4 light years away. Sirius, most brilliant of the night sky's stars, is 8 light years distant.

◄ The central region of the Orion Nebula, including the bright Trapezium multiple star system at its core.

▶ The Crab Nebula is the remnant of a supernova which exploded in Taurus in 1054.

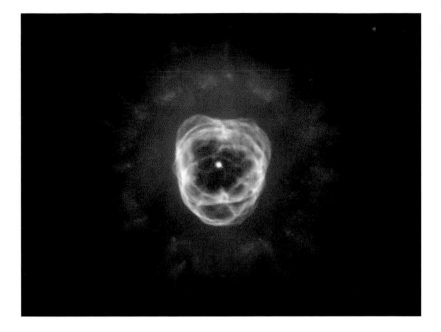

These stars, like our Sun, are part of the home Galaxy, the Milky Way – a flattened disk containing an estimated 200 billion stars, together with assorted dust, gas and other material, spanning a total diameter of over 100,000 light years. Our Galaxy (the capital "G" is to distinguish ours from the countless other galaxies astronomers can detect in the Universe) has a spiral structure, and from our vantage point in one of the outer, star-rich spiral arms about 25,000 light years from the central bulge we can see bright nebulae – clouds of gas and dust in which new stars are forming – and clusters of gravitationally bound, recently formed stars. The classic nebula, known to many with only a casual interest in

▲ The Hubble Space Telescope image of the Eskimo Nebula in Gemini shows intricate detail shaped by the central star's magnetic field. This is a planetary nebula, formed at a late stage in the life of a Sun-like star.

astronomy, is the Orion Nebula (M42), which lies 1600 light years away and is faintly visible to the naked eye on a good, dark clear night. (The Eagle Nebula is 6500 light years away, and rather fainter.)

The naked eye will also easily show the famous Pleiades (Seven Sisters) star cluster in Taurus, a compact grouping of hot, young stars 410 light years away. Binoculars and small telescopes reveal many other nebulae and clusters in our Galactic neighborhood out to considerable distances.

Among the other denizens of the Galaxy are planetary nebulae, produced when aging giant stars shed their outer layers; and supernova remnants, which are the shattered and scattered leftovers from the explosion of very massive stars which have reached the end of their nuclear-fusion lifetimes.

Our Galaxy has a central bulge, surrounded by a more distant spherical "halo" of globular clusters – dense collections in which hundreds of thousands of stars may be compacted into a volume of space typically 200 light years across. Some of our Galaxy's more remote globular clusters are as much as 200,000 light years away. Some of the less distant, brighter examples are well within binocular

▲ *Omega Centauri, the brightest globular cluster associated with our* *Galaxy, is a spectacular object for observers at southerly latitudes.*

► *The Hubble Deep Fields, very long exposures taken with the orbiting telescope in 1995 and 1998, reveal extremely distant galaxies formed early in the history of the Universe.*

range; indeed, observers in southerly latitudes can see two of the brightest globular clusters (Omega Centauri and 47 Tucanae) as fuzzy "stars" with the naked eye.

Beyond the halo of the Milky Way, we enter the realm of truly vast distances, in pursuit of other galaxies. Familiar to southern observers are the Magellanic Clouds, dwarf satellite galaxies orbiting the Milky Way at respective distances of 170,000 and 190,000 light years. Further afield, 2.4 million light years away, is the Andromeda Galaxy (M31), visible with the naked eye to observers in the northern hemisphere on dark, transparent autumn evenings.

Galaxies tend to come in clusters. The Andromeda Galaxy and the Milky Way are the two largest members of a relatively small cluster known as the Local Group. Galaxies belonging to other clusters can be picked out in quite modest telescopes and even binoculars. For these objects our lookback time is considerable: for example, many of the galaxies in the Virgo–Coma Cluster (named for the constellations in which they are seen) lie 65 million light years away – the light from these galaxies arriving in our telescopes today was emitted at the end of the dinosaurs' reign on Earth.

On the cosmological scale the Virgo–Coma Cluster is quite nearby; most other galaxy clusters are even more distant. Amateur telescopes can detect galaxies at incomprehensibly vast distances but cannot, naturally, compete with the lookback time of such images as the Hubble

▲ *The center of our Galaxy lies in the direction of Sagittarius. Here the Milky* *Way broadens and appears at its brightest. Dark dust lanes are also visible.*

Deep Fields of 1995 and 1998, which offer a view of then recently formed galaxies from 10 billion years ago, only a couple of billion years after the Universe's creation in the Big Bang.

Some astronomers talk of deep sky observing as the capture of "ancient light." Certainly, alongside the practical challenge and personal achievement of locating faint nebulae, clusters or galaxies, there is a philosophical edge to deep sky observing when one considers the scale and age of the Universe revealed in the eyepiece.

### Seasons in the Sky

Our view of the deep sky varies with the seasons. Some classes of object are better seen at certain times of year than at others, partly as a result of the Earth's location in one of the Galaxy's spiral arms. The Milky Way, which from dark observing locations can be seen spanning the midnight sky between June and August, is the combined light of millions of stars in spiral arms, and is brightest along our line of sight toward the Galactic hub, which appears as a broadening of the Milky Way containing numerous dense starclouds in the constellation Sagittarius.

Objects are best seen when at their highest above the horizon, culminating on the meridian (p.22). From June to August the midnight meridian is dominated by objects close to the Milky Way's relatively flat plane, making this a prime time of year to observe the open star clusters and star-forming nebulae that lie along the Galactic spiral arms. Many of the globular clusters in the halo surrounding the Galactic center are also on view.

As the Earth continues in its orbit, the midnight vista gradually slips eastward from day to day so that by September and October the view is away from the plane of the Milky Way and out into the real depths of intergalactic space. This is a good time to look for galaxies. Few such objects are visible in the direction of our own Galaxy's heart, because our view into intergalactic space is obscured by gas, dust and stars (professional astronomers at one time spoke of the Milky Way plane as a "zone of avoidance" for distant galaxies).

By December, the Sun has moved into Sagittarius, and objects in this region are lost in its glare. Meanwhile, the midnight view is toward the next spiral arm out from our own, in the direction of the bright stars in Orion and the surrounding constellations. The Orion Nebula is the brightest part of a great complex of star-forming gas and dust in this region. While fainter than the stretch of Milky Way visible in the middle months of the year, the section on view from December to February is home to numerous open star clusters along its length.

▶ Our Galaxy is a spiral galaxy, perhaps similar to this galaxy, M74, which is seen face-on. The star-rich central bulge is surrounded by several wound-up spiral arms, in which "knotted" structure indicates regions of star formation.

March and April bring deeper views again, and are good months for trawling the Virgo–Coma Cluster of galaxies. May and June offer good views of globular clusters in Ophiuchus and Scorpius.

## An Amateur Astronomer's View of the Eagle Nebula

The Eagle Nebula (M16) is one of the prominent objects in the part of the Milky Way best seen in the late evening between about June and August. Found southwest of Aquila in the relatively obscure constellation of Serpens, M16 is an easy target for binoculars and small telescopes, but what they will show the amateur observer is nothing like the familiar Hubble Space Telescope portrait. M16 was first described in 1746 by the Swiss astronomer de Chésaux (best remembered for discovering the bright multitailed comet of 1743). His telescopic view was very similar to what modern binoculars reveal – a hazy patch, elongated east–west, in the rich Milky Way starfields.

What binoculars show is actually the combined light of the Eagle Nebula and the star cluster NGC 6611 associated with it. These stars are nicely resolved in a small telescope at modest magnification.

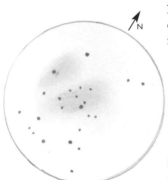

At a magnification of ×20, my 80 mm aperture wide-field refractor shows the cluster as a compact scattering of stars; increasing the magnification to ×40 fills about half the field with faint stars. However, the nebulosity that is so distinctive in long-exposure photographs taken with large telescopes is visually elusive in small telescopes, showing only as a faint background haze. The nebulosity is cataloged separately from the cluster as IC 4703, and was first mentioned in descriptions of the object by the French comet-hunter Charles Messier in 1766.

▲ The Eagle Nebula as sketched by the author observing with his 80 mm refractor at ×40. This small amateur telescope view naturally reveals much less detail than that from the Hubble Space Telescope. Visible here are parts of the nebulosity IC 4703 associated with the star cluster NGC 6611, which together comprise M16.

While they may appear dim, it is worth remembering that the stars in NGC 6611 are intrinsically much more massive and luminous than our Sun; they appear faint only by virtue of their immense distance.

## About the Book

M16 is one of thousands of deep sky objects which can be found with equipment readily available to amateur astronomers. Many more will be highlighted later in this book, but first

– in the next chapter – I shall take a look at a few basics regarding how to go about observing them, and the capabilities of typical amateur binoculars and telescopes for deep sky observing.

The chapters that follow offer a selection of deep sky objects suitable for small to medium amateur instruments; some are even visible to the naked eye. I have arranged the objects in each chapter roughly in order of increasing difficulty; sometimes, though, it is more sensible to discuss easy and more difficult targets together if they are close in the sky. In each chapter a few more testing objects are also included. I start with galaxies and their putative building-blocks, the globular clusters. The chapters then follow a rough stellar-evolution sequence from diffuse nebulae, through star clusters and double stars, to planetary nebulae, which mark the end of the nuclear lifetime of a Sun-like star. The most violent stellar ends are supernova explosions, and the remnants of a few of these catastrophic events are visible in amateur telescopes.

The final chapters look at how deep sky converts might take their interest a stage further, and present a brief outline of the historical development of deep sky observing. At the end of the book are lists of deep sky objects for quick reference; sources of information, both printed and on-line; and all-sky charts (more detailed charts to help in the location of specific objects are included in the relevant chapters).

# 2 · THE BASICS

An obvious first question facing the would-be deep sky observer is just what sort of equipment is required to bring targets into view. Several deep sky objects are visible without optical aid: in the northern hemisphere lie the Hyades and Pleiades star clusters and the Andromeda Galaxy (M31); in the southern hemisphere, the Magellanic Clouds and the globular clusters Omega Centauri and 47 Tucanae are familiar naked-eye sights. These are all quite bright, but to reveal most deep sky objects the observer will need the additional light-gathering power of at least binoculars or a small telescope.

Instruments and accessories, and what I think are the best choices for the beginning deep sky observer, are discussed in more detail later. First, though, it is worth discussing how one finds one's way around the sky.

## Getting around the Sky
### The Magnitude Scale

A good starting point in appreciating the instrumental requirements for deep sky observing is to look at the magnitude (brightness) scale used by astronomers. Historically, the scale goes back to the second century BC and the star catalog drawn up by the Greek astronomer Hipparchus of Nicea (c.190–120 BC). Hipparchus assigned each star to one of six magnitude classes, the brightest being of first magnitude, those slightly fainter of second magnitude, and so on down to the naked-eye limit at sixth magnitude. Thus the greater an object's magnitude, the fainter it is. In the nineteenth century astronomers placed the scale on a firmer mathematical footing by defining a difference of one magnitude as equivalent to a factor of 2.512 in brightness; a difference of five magnitudes is thus a 100-fold difference in brightness, so a sixth-magnitude star is 100 times fainter than one of first magnitude.

The bright star Vega in Lyra was used as a standard in establishing modern magnitude scales, and was assigned a value of 0.0. A few stars are brighter than Vega and have negative magnitudes – the brightest is Sirius, at mag. −1.47. The brighter planets can also have negative magnitudes, typically −2 for Jupiter and −4 for Venus (their brightness varies with their changing distance from Earth). Objects fainter than magnitude 0 often have a "+" added before the value to remove any ambiguity – for example, the northern pole star, Polaris, is of mag. +2.0; if there is no "+," the magnitude can be assumed to be positive.

Deep sky objects are typically below the naked-eye limit. One of the brightest, the Andromeda Galaxy, is a fairly faint naked-eye object at mag. +3.4. Good objects for observing with binoculars or a small telescope are around mag. +6 to +8; most are fainter. As we shall see, the ability to detect faint objects depends in large part on the diameter of the instrument's light-collecting lens or mirror.

Another factor to bear in mind is that many deep sky objects have their light spread over an extended area of sky. Catalogs of deep sky objects usually list an overall, *integrated magnitude* for each one, but for extended objects the surface brightness, and therefore contrast with the background sky, will be low. A planetary nebula with a catalog magnitude of +9, for example, will be harder to see than the sharp, point source of a 9th-magnitude star.

### Sizes and Distances – the Angular Scale

Together with its brightness, the apparent size of a deep sky object has an important bearing on the choice of instrument used to observe it. Some of the relatively nearby open star clusters – like the Pleiades in Taurus – appear quite large on the sky, and are best observed with binoculars or a wide-field telescope at low magnification. On the other hand, many planetary nebulae or distant galaxies may present such small apparent diameters that only a high magnification on a telescope will reveal anything of their true nature.

Sizes and distances on the sky are measured as angles. For convenience, astronomers have long dealt with sizes, positions and motions on the sky in terms of an imaginary celestial sphere surrounding the Earth, onto which the stars and other objects appear to be projected – as if onto the inside of a vast planetarium dome. The celestial sphere has, of course, a circumference of 360 degrees. A degree (°) on the sky subtends a distance roughly twice the apparent diameter of the full Moon. Each degree can be subdivided into 60 arcminutes ('); the Moon's approximate half-degree diameter can also be expressed as about 30 arcminutes. In turn, each arcminute is made up of 60 arcseconds (").

In terms of these angular sizes, the Pleiades have a diameter of about 2° – too large to be contained in a single high-magnification telescopic field, but comfortably accommodated in the field of view of, say, 10 × 50 binoculars. The Eagle Nebula spans 35' × 28' – a good-sized object for a modestly powerful telescope. Some planetary nebulae, on the other hand, are rather compact: the Eskimo Nebula (NGC 2392) in Gemini, for example, has a small, 15-arcsecond disk and so is best seen in a narrow, relatively high-magnification telescopic field of view.

The angular scale is also useful for describing distances between objects. The components of a double star system might be described as being 15 arcseconds apart, for example, while the open star clusters M46 and M47 in the constellation Puppis are separated by 1.5 degrees (1° 30'), or roughly three Moon-widths.

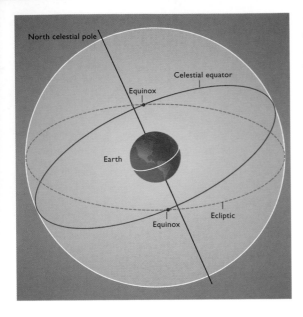

◀ The imaginary celestial sphere surrounding the Earth is a useful means of visualizing positions on the sky. Projected onto the sphere are the Earth's poles and equator. The dashed line of the ecliptic marks the apparent path of the Sun against the star background over the year. The ecliptic is inclined to the equator, intersecting it at the equinoxes.

### Directions and Angles

As well as being familiar with angular measures, the observer trying to locate deep sky objects needs to know the conventions used for directions in the sky. For example, one might see the Andromeda Galaxy (M31) described as being located about a degree to the west of the fourth-magnitude star Nu Andromedae. Such directional information can be understood by thinking about how the sky appears to move as a result of the Earth's daily rotation. Objects are carried westward across the sky as the Earth rotates. For an observer in the northern hemisphere, west is to the right and east to the left; M31 therefore appears a couple of Moon-widths to the right of Nu Andromedae.

The directions of north and south are indicated by the two celestial poles, around which the stars appear to wheel. For northern-hemisphere observers, the position of the north celestial pole is indicated quite well by the second-magnitude star Polaris: north is more or less in the direction toward Polaris, and south is the direction away from Polaris. For objects on the meridian to the south of the sky, north is uppermost in the naked-eye view, while south is in the direction of the horizon. In the southern hemisphere, west is to the left and east to the right, while south is in the direction of the south celestial pole (for which there is no convenient, prominent marker star).

Sometimes, to avoid confusion, astronomers use the terms preceding (*p*) and following (*f*) instead of west and east, respectively. This is

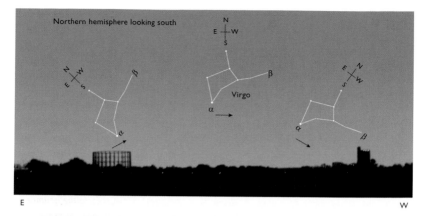

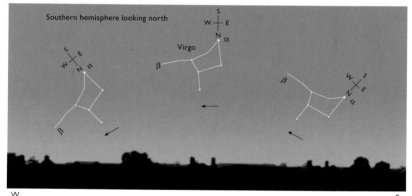

▲ *Motion and orientation in the sky. As a result of the Earth's rotation, stars move westward across the sky in the course of the night. In the northern hemisphere (top), the stars of the constellation Virgo, for example, rise in the east, culminate (reach their highest) due south, then set in the west. They move steadily from left to right for an observer facing south. In the southern hemisphere (bottom), however, culmination occurs due north and motion is from right to left for an observer facing north; also, the stars appear inverted relative to the northern hemisphere view.*

common practice among planetary observers, but it is equally appropriate to describe, say, a globular cluster as being "40 arcminutes south preceding" (abbreviated S*p*) a field star.

Another useful way of describing relative location is by *position angle*. One can visualize a clock-face centered on a star, for example, with the cardinal points (north, east, south and west) around its edge. The direction in which another object lies relative to the star is indicated by

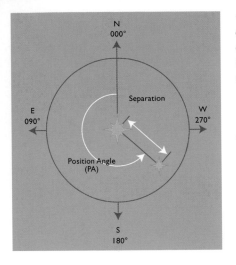

◄ *Position angle (PA) is a useful means of indicating the relative location of objects. PA is measured in degrees from 000 due north, through 090 due east, and so on.*

an angle between 0 and 360 degrees, reckoned counterclockwise from north. An object due north of the star is said to be at position angle (PA) 000°; due east, 090°; due south, 180°; and due west, 270°. For example, M1, the Crab Nebula in Taurus, can be described as lying a degree from the third-magnitude star Zeta Tauri at PA 320°.

Position angle is also used to indicate the relative positions of the members of a double star system (see Chapter 7). The position angle of the secondary (usually fainter) member relative to the primary is normally given in listings of double stars, together with the stars' separation in arcseconds. Another useful application of position angle is in defining the long axis of, say, an elongated, edge-on galaxy: NGC 4565 in Coma Berenices has its long axis in PA 135–315°, for example.

## Positions on the Sky

For the successful detection of deep sky objects, just as important as an understanding of the magnitude scale is a good working appreciation of positions and scale on the sky. Navigating one's way around the stellar heavens can be a daunting prospect at first, but with practice it becomes second nature. Observers who choose to use the increasingly common "GOTO" telescope mountings, which incorporate a computer that can automatically point the instrument at a preprogrammed target, may enjoy the convenience of never having to really learn their way around the sky. For the traditionalist, though, part of the thrill of the chase is using one's own skill and knowledge of the telescope's capabilities to hunt down a faint quarry.

### Equatorial coordinates

Positions of objects on the sky are defined in different ways. Most fundamental is the equatorial system of right ascension (RA) and declination (dec), the coordinates that on the celestial sphere correspond respectively to terrestrial longitude and latitude. Declination is measured north (+) or south (−) of the celestial equator, a projection onto the celestial sphere of the Earth's equator. Declination is measured in the angular units of degrees, minutes and seconds of arc, already discussed. Right ascension is measured eastward from the point where the ecliptic – the Sun's apparent annual path against the star background, inclined at an angle of 23.5° to the celestial equator – crosses the equator heading northward, at the spring equinox. Known, for historical reasons, as the First Point of Aries, this position currently resides in the constellation Pisces. RA is most commonly measured in hours, minutes and seconds. Each hour of RA spans 15° on the celestial sphere.

The standard grid that forms the basis of the equatorial coordinate system shifts gradually relative to the stars as a result of the roughly 26,000-year precession ("wobble") of Earth's axis of rotation. Over time, the northern spring equinox marking RA 0h 00.0m slips slowly westward through the zodiacal constellations. Consequently, astronomers have to periodically revise positions in catalogs of stars and other celestial objects. This is done at 50-year intervals, so that positions are defined for standard *epochs*. In this book, positions refer to the epoch of 2000.0. The inexorable progress of precession will lead to a revision of positions such that these will, later in the twenty-first century, be referred to the epoch of 2050.0.

As an example of the use of equatorial coordinates, the Eagle Nebula M16 has epoch 2000.0 coordinates of RA 18h 18.8m, dec −13° 47′.

► Equatorial coordinates on the celestial sphere. Right ascension (RA), the equivalent of longitude, is measured eastward from the northern hemisphere spring equinox point (see also p. 18). Declination, equivalent to latitude, is measured north (+) or south (−) of the celestial equator.

### Horizontal coordinates

Out under the real sky, the most convenient way for the observer to think about objects' positions is in terms of their altitude and azimuth – in horizontal coordinates. An object's altitude (or elevation) is measured in degrees, minutes and seconds above the horizon. Azimuth (bearing) is measured westward from due north at 000 degrees, through due east (090 degrees), due south (180 degrees), due west (270 degrees) and back round to 000/360 degrees at due north again.

Horizontal coordinates relate in a couple of important ways to the equatorial system. Objects will be at their highest above the horizon when due south for an observer in the northern hemisphere, or due north in the southern hemisphere (in other words, on the equatorward meridian). The meridian is an arc of a great circle passing through the north and south celestial poles, stretching from the horizon to the zenith point overhead. Objects rise in the east, reach their highest point (culminate) on the meridian and set in the west. In the northern hemisphere, objects with a high "+" declination will culminate further above the horizon than those with lower, or negative declinations, while the reverse applies in the southern hemisphere.

It is also now easy to see why right ascension is measured in hours. The line of RA on the meridian slowly increases as the Earth rotates, so that, for example, if RA 21h 36m is on the meridian at a given time of night, then, one hour of time later, the line of RA on the meridian will be 22h 36m.

As a result of Earth's orbital progress around the Sun, and the fact that the day is 23h 56m 4s long (rather than exactly 24 hours), the line of RA on the meridian at the same clock time moves by roughly 4 minutes per day. Objects on the meridian at midnight at the end of any month will be about 2 hours of RA east of objects that occupied the midnight meridian at the start of the month. This is what causes the gradual seasonal shift in the parts of the sky on view. Orion and its neighboring bright constellations are on best evening view from December to February (winter in the northern hemisphere, summer in the southern), while Sagittarius with its rich Milky Way starclouds is best seen in June and July (northern summer, southern winter).

### How low can you go?

What objects are visible over the course of a year depends on the observer's latitude. An observing site at high northerly latitudes will always be denied access to objects such as the Magellanic Clouds which have high southerly declinations, while observers in central Australia will find the Ursa Major galaxy pair M81 and M82 permanently below their horizon. How far down in declination one can reach from latitude $L$ may be determined from $-(90-L)°$. For example, from my home in southern England at latitude 51°N, I can – just! – see down to a declination of $-(90-51)° = -39°$, and the "Sting" stars of Scorpius (Lambda and Upsilon Scorpii), at dec $-37°$, scrape low over my southern horizon on a midsummer's night. For observers in Sydney, Australia (latitude 34°S), objects north of dec +56° are inaccessible.

Objects at declinations just above the lowest limit visible from the observer's latitude will be poorly seen, low over the southern horizon in the northern hemisphere (the northern horizon in the southern hemisphere). Particularly within about 15° of the horizon, the effect of atmospheric extinction takes a significant toll on the apparent brightness of objects: the farther an object is from the zenith, the point directly overhead, the greater the thickness of atmosphere the object's light passes through on its way to the observer. The long "wedge" of air through which a low target is viewed serves to attenuate its light. The Sting stars in Scorpius, for example, are of second magnitude, in the brighter end of the naked-eye range. However, at just 5° above the horizon, almost two magnitudes of light are lost to atmospheric extinction, and in order to see Scorpius' Sting well at my location I have to use binoculars. The effect of extinction on extended objects such as nebulae is far worse.

As a general rule, it is best – at least for the novice observer – to concentrate on deep sky targets which culminate at an altitude of 15° or greater. In other words, it is best to look for objects whose declination is 15° north or south of the absolute visibility limit at northern- and southern-hemisphere locations, respectively. If a pressing desire to seek out objects very low down does take hold, bear in mind that the view will be compromised by atmospheric extinction (and will be made worse if conditions are hazy).

## Equipment and Accessories
### The Eye

Not to be overlooked is the most fundamental piece of observing equipment – the "Mark 1 eyeball." Even without optical aid, the eye can detect some faint objects under the right conditions. Although most deep sky objects will be observed with binoculars or a telescope, the eye is still a crucial part of the optical system. In trying to detect faint objects at night, the eye's physiological response to dark conditions comes into play: this is the phenomenon of *dark adaptation*.

On first going outdoors from a brightly lit room on a clear, dark night, you may struggle to see much more than the brighter stars – down to fourth magnitude, perhaps. After a while you will become aware of fainter stars, and after 20 to 30 minutes under dark conditions you should be able to see stars to sixth magnitude.

The physiological response leading to dark adaptation is twofold. First, the pupil of the eye dilates, opening to make the lens a larger light-collecting area. More important, however, are the slower changes in the light-sensitive retina at the back of the eye, which include production of the pigment rhodopsin ("visual purple"), which aids night vision, and the use of rod cells as the dominant receptors (as opposed to the color-sensitive cones which dominate in bright-light conditions – see below).

It has been suggested that observers will adapt to the dark more quickly if they wear sunglasses during the day ahead of a session with the telescope. Certainly, it is wise to avoid any very bright light source just before going out to observe; I find that it takes considerably longer for my night vision to become established if I have been working at the computer screen, for example.

Experienced deep sky observers use a simple technique called *averted vision* to help them detect faint objects with the dark-adapted eye. To understand how this works, it is first worth looking more closely at the surface of the retina. It contains two types of receptor cell: light-sensitive rods (about 120 million of them) and color-sensitive cones (6–7 million). The cones are concentrated in a central, yellow spot on the retina called the macula, densely packed into a tiny (0.3 mm wide) area known as the fovea centralis. Rods are distributed in the peripheral regions, and are at least a thousand times as light-sensitive as the cones. It therefore helps to look slightly to one side of the object in the eyepiece, so that its light falls on the rods rather than on the less-sensitive cones. Although there is some variation from one observer to the next, the best results are usually obtained by having the object more toward your nose in the field of view. In other words, if, like mine, your right is the "master" eye, look slightly to the right of the object in order to bring its light onto the rods in that eye.

The night-vision rods give only weak color perception, shifted to shorter wavelengths than in the daylight color vision provided by the cones. Red light is relatively poorly detected, and faint objects viewed in the eyepiece will typically appear white or gray, though some planetary nebulae may look blue or greenish. The observer certainly shouldn't expect to see the sometimes garish colors of images from, for example, the Hubble Space Telescope which are reproduced in the popular literature: much of the color in these images results from long exposures and subsequent processing which enhances features by exaggerating color contrast.

### Protecting night vision

Once your eyes have become dark-adapted, you'll want to keep them that way for the duration of the observing session. Deep sky specialists seek out the least light-polluted areas; in some urban locations, it is impossible to gain full dark adaptation.

There is the problem that while observing you will wish to take notes, consult charts and tables, make drawings, or change eyepieces on the telescope – all activities that require some level of light. Use of a strong white light is very much to be avoided, as this will destroy your night vision. The solution is to use a dim red light, as red light has the least detrimental effect on night vision. A number of equipment suppliers sell LED-based lights which work well. Alternatively, you can do as I have, and take an off-the-shelf flashlight and cover the glass at the front with overlapping light-tight strips of red insulating tape.

Some observers go to remarkable lengths to cut out extraneous light when hunting down faint objects, even avoiding use of red light during the session. It has been known for observers to use a black hood or sheet over their head and the eyepiece-end of the telescope, in the fashion of old-style photographers, to exclude completely any distracting light sources which might interfere with target detection.

### Binoculars for Deep Sky Observing

A common misconception among newcomers to astronomy is that a telescope – preferably a large one – is essential if any worthwhile views are to be had. Nothing could be further from the truth: I know very experienced observers who make valuable studies of variable stars, for example, equipped with no more than a pair of hand-held binoculars. Such instruments can also be put to good use for hunting out deep sky objects.

Binoculars have the advantages of relative cheapness and portability. They are described, usefully, in terms of their magnification and the aperture of their front-end – *objective* – light-gathering

▶ Binoculars are ideal for wide-field deep sky viewing. Mounting them on a tripod allows more comfortable observing, and is essential for larger, heavier instruments.

lenses. For example, the binoculars I have used for a lot of observing in the past 28 years are described as $10 \times 50$ – they have objectives of 50 mm aperture and magnify 10 times ($\times 10$). These give me comfortable access to stars down to about magnitude +9.5 from my semirural observing location, and show reasonably faint deep sky objects such as the 8th-magnitude Crab Nebula.

The aperture of the objective (or primary mirror in a reflecting telescope) is a critical factor in an instrument's light-gathering capability. Within the limits imposed by local observing conditions, bigger is better in terms of light-gathering, so a 50 mm objective will outperform a 40 mm in terms of bringing faint objects within reach, while 70 or 80 mm is better still. The combination of twin objective lenses in binoculars brings its own advantages: two eyes are better than one, so 50 mm binoculars will allow access to fainter objects than a single 50 mm lens.

Some observers prefer lower-magnification $7 \times 50$ binoculars. The cone of light emerging from each eyepiece is wider and brighter than in $10 \times 50$s. (The maximum cross-section through the cone of light is the *exit pupil*, found by dividing the aperture by the magnification: 5 mm for $10 \times 50$s, about 7 mm for $7 \times 50$s.) This has some advantages for younger observers, whose fully dilated pupils will be able to take in the full beam. Older observers, however, gain little from this as the pupils' maximum dilation diminishes with age.

A pair of $10 \times 50$ binoculars can be comfortably hand-held for maybe a minute or two, but after that the strain of holding them steady becomes noticeable. The strain can be relieved by bracing one's arms on a convenient wall. For viewing objects overhead, lying back in a lounger, using the arms for support, is a good idea. An adapter which allows the binoculars to be attached to a camera tripod is a worthwhile investment.

Larger-aperture binoculars are available, and have become popular with deep sky observers. Common formats include $15 \times 70$ ($4.4°$ field) and $20 \times 80$ ($3.3°$). These larger instruments cost more, of course, and will very definitely need to be mounted on a tripod. Also expensive are image-stabilized binoculars, engineered with movable optics to counter hand-shake. The necessary internal gadgetry adds to the weight of such instruments, and in the end the deep sky observer is probably better off using standard binoculars on a tripod. At the top end of the scale are true giants, such as ex-military $25 \times 100$ binoculars ($3°$ field) which offer stunning views, but are very expensive and are most suited to the specialist observer.

The wide field provided by binoculars shows some objects to good advantage. For instance, $10 \times 50$s offer a $5°$ field which will show the Pleiades in their entirety; $7 \times 50$s give a $7°$ field. Binoculars will allow access to dozens of open star clusters. Brighter globular clusters will

▶ Optical layouts for common telescope systems. In a simple refractor, light collected by the front-end objective is brought to a focus at the other end of the tube. In a Newtonian reflector, light is gathered by a bottom-end mirror, and reflected via a secondary to focus at the side of the tube near the top. Catadioptrics use a combination of a primary mirror and a curved transparent correcting plate on which the secondary mirror is a silvered spot. Light is brought to a focus through a hole in the primary mirror at the bottom end.

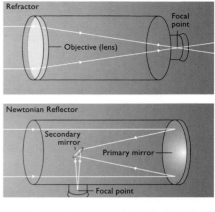

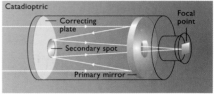

appear as fuzzy "stars," as will some galaxies. Some diffuse nebulae show well (the Orion Nebula being the obvious example), but with the exception of the Dumbbell (M27), planetary nebulae are too small to be seen to advantage. Quite likely, the binocular deep sky observer will eventually wish to move up to an instrument with a bit more magnifying power. Even when using a telescope, however, binoculars remain a useful aid: I find it helpful when attempting to locate fainter objects with a telescope to first make a quick reconnaissance of the field at low power using 10 × 50s.

▼ A small aperture, short-focus refractor like this is an ideal portable instrument which can be mounted on a camera tripod for convenient viewing. The smaller telescope mounted on top is a finder.

## Telescopes for Deep Sky Observing

Binoculars offer good, wide-field, low-magnification views of the sky, but for higher-powered viewing a telescope is a must. Three basic telescope types are widely available off the shelf from commercial suppliers: refracting telescopes (also called refractors), reflecting telescopes (reflectors) and

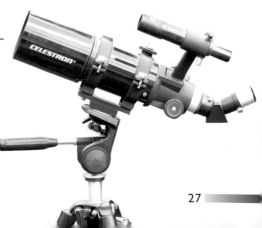

catadioptric (or Schmidt–Cassegrain) telescopes. As with binoculars, the clear aperture of the light-collecting objective lens in a refractor, or the primary mirror in a reflector, is critical: a large-aperture telescope will show fainter objects.

### Refracting telescopes

Refractors use a front-end objective lens to gather light and bring it to a focus in the eyepiece at the other end of a straight tube. Modern refractors are of two kinds. Short focal length instruments giving wide-field views are popular with many observers (including the author), and have the advantage of portability. Cheaper short-focus refractors are notoriously prone to *chromatic aberration*, where light of blue wavelengths is brought to a slightly different focus than light of red wavelengths, leading to spurious colored fringes around the images of brighter objects, particularly when high magnifications are used. Double-element – *achromatic* – objective lenses, which combine two different types of glass, help to counter chromatic aberration, as can the use of color-correcting eyepiece filters. It has to be said, however, that chromatic aberration is more of a problem when observing bright, planetary targets than it is for most deep sky objects. A number of manufacturers now produce very high quality short-focus instruments with triplet objectives made from ED or EHD glass, coupled with improved surface coatings on the optics. Such *apochromatic* telescopes are superb traveling companions for those who can afford them.

◄▲ *Long-focus refractors are better for high-magnification viewing, but may entail some awkward viewing positions unless a star diagonal (inset) is used. The telescope is on an equatorial mount, which allows the stars to be followed across the sky more easily.*

Long focal length refractors are less prone to chromatic aberration, and again apochromatic versions can be obtained at a price. If higher magnifications are the observer's desire (for detailed views of planetary nebulae, say), a long-focus refractor is ideal. The downside to refractors, and long-focus/long-tube models in particular, is that the observer can end up in some very uncomfortable positions when trying to view objects high in the sky: the straight-through view takes the eyepiece position close to the ground on most standard mountings once a refractor is aimed above about 50° elevation. One solution is to employ a *star diagonal* – a mirror housing that redirects the light path 90° to the tube for more comfortable viewing. The principal disadvantage of star diagonals is that they give a laterally reversed, mirror image of the sky, which can be disorienting for the observer. It is very much a matter of personal choice whether to use a diagonal or to persevere with a cricked neck!

### Reflecting telescopes

Classic-design *Newtonian reflectors* use an optically figured large *primary mirror* at the bottom of the tube to collect light, which is then directed via a smaller secondary mirror to the eyepiece mount/focuser on the side of the tube, usually near the top end. Viewing objects at high elevations is therefore less of a problem with Newtonians than with refractors. For any given aperture, a mirror is cheaper to make than an objective lens: a 150 mm aperture reflector is a good deal less expensive (and a lot less unwieldy) than a 150 mm aperture long-focus refractor. Since light of all wavelengths is reflected equally, chromatic aberration doesn't afflict reflectors (unless it is introduced by using poor eye-

▶ An equatorially mounted Newtonian reflector, showing the focusing mount on the side of the tube near the top end. This arrangement allows more comfortable viewing of objects at high elevations.

29

pieces). The assembly that holds the secondary mirror in a reflector does obstruct the light path somewhat and reduce the image contrast, but most observers don't find this a particular problem for deep sky objects.

Newtonian reflectors do require some maintenance. In particular, the observer has to be careful to ensure that the optics are kept in good alignment – that they are properly *collimated*. A poorly collimated reflector will produce distorted, non-point star images. Alignment can often be lost during transit if a reflector is transported to an observing site, and some extra preparation will then be necessary to recollimate the optics before observing can begin. For some observers, freedom from regular recollimation makes refractors preferable as portable instruments.

*▼ Catadioptric systems have become popular in the last 20 years or so. This model, on a fork mount, is equipped with a GOTO facility, which is operated via the handset.*

### Catadioptric telescopes

Schmidt–Cassegrain (SCT) or catadioptric systems, which are particularly popular in the United States, combine mirrors with smaller refracting elements, compressing the light path to fit into a comparatively short (and therefore more portable) tube assembly. Light is brought to a focus through a central hole in the primary mirror, and it is usual to have a laterally reversed view in the eyepiece via a diagonal: as in refractors, a straight-through view makes it inconvenient for the observer when the target is at a high altitude. As with Newtonian reflectors, maintaining collimation is important, but this is easier to achieve with SCTs, which have sealed optical assemblies.

Choice of instrument is largely down to observer preference and circumstances. In terms of portability, 7 × 50 or 10 × 50 binoculars are hard to beat, but these provide only low-magnification views. Slightly more powerful 15 × 70 or 15 × 80 models gain a little in magnification and light grasp, but will be weightier and require tripod-mounting.

Short-focus refractors are also fairly portable, but budget models can suffer from chromatic aberration and are less able to deliver quality, high-magnification views than their long-focus counterparts. Long-tube refractors, while less portable, offer good high-magnification images (making them suitable for planetary as well as deep sky observing) but also lead to some awkward viewing positions for objects high in the sky unless a diagonal is employed. Newtonian reflectors are cheaper, aperture for aperture, than the equivalent refractors, but need more careful maintenance to keep the optics properly aligned. Long focal-length reflectors can be cumbersome from the point of view of storage, but the popular SCT designs with their "folded" optics are more compact.

## Telescope Mountings

An astronomical telescope – whatever its type – is only as effective as its mounting. A telescope on an insufficiently substantial or stable mount will be difficult to use and a source of considerable frustration. The ideal mount will allow the telescope to be moved smoothly, and have minimal shake: vibrations after adjusting the focus, for example, should preferably settle down after no more than a couple of seconds.

Large telescopes need large mountings, but some of the lighter and more portable models can be mounted satisfactorily on a good camera tripod. My short-tube 80 mm refractor has a mount with a bushing so that it can be attached to a camera tripod, and takes just moments to set up. The simple camera tripod is an example of an *altazimuth mounting*. In other words, the telescope can be freely moved up and down (in altitude) and from side to side (in azimuth). This is perfectly adequate for a small, wide-field refractor.

▶ The altazimuth Dobsonian mounting is cheap and effective for Newtonian reflectors.

A novel form of altazimuth mount was introduced by the American observer John Dobson. Simplicity and low cost are its main attractions, and since the 1980s thousands of amateur astronomers around the world have constructed *Dobsonian mountings* for their reflectors. Many commercial equipment makers also supply Dobsonian mounted reflectors. Simply described, the Dobsonian mounting achieves movement in altitude and azimuth through a so-called rocker box which holds the bottom – primary mirror – end of the tube. The assembly may be set up on flat ground or on a low table, the mount taking its stability from the telescope's low center of gravity. The simple bearings are coated with Teflon, allowing the instrument to be moved smoothly from one position to the next.

While simple altazimuth mounts are fine for small refractors, and Dobsonians are excellent for reflectors even up to quite large apertures, larger instruments are more usually set up on *equatorial mountings*. The traditional *German equatorial* is a proven design which has been popular with amateur astronomers for a very long time. An equatorial allows motion around two axes. One – the *polar axis* – is aligned with the celestial pole. The telescope can be driven around the polar axis by a motor to counter the effects of Earth's rotation: tracking at the same pace as the stars' apparent motion (the *sidereal rate*) is a convenience for the visual observer, and essential for those aiming to take long-exposure images through the telescope (who will also need precise alignment of the polar axis with the celestial pole). The telescope can also be moved up and down – north and south – about the other axis, the *declination axis*. Many basic mountings have simple manual slow-motion controls which allow the telescope to be nudged along a little at a time, and are adequate for most visual observing.

Schmidt–Cassegrain telescopes are commonly provided on an equatorial called a *fork mount*, with the base of the assembly tilted to be perpendicular to the celestial pole. The fork is on the polar axis, and the telescope is free to move up and down in declination. Smaller SCTs, and medium-sized telescopes on German equatorial mounts, are reasonably portable. For observers who become seriously interested in imaging work and/or wish to move up to larger instruments, the eventual solution may be to place the telescope on a permanent mounting (removing the need to align the polar axis before each session), and perhaps even to house it in its own observatory.

Recent years have seen the introduction, particularly for SCTs such as Meade's popular ETX range of telescopes, of "smart" mounts which can be set up in essentially altazimuth mode and programmed to slew to objects whose positions are stored in an onboard computer. At the start of the observing session, a couple of reference stars have to be located to prime the mount, which thereafter can find other objects from their angular offset.

The convenience of such GOTO technology is obvious, but its use has been the subject of heated debate at local astronomy clubs and star parties. Purists argue that observers who use automatic object location have no need to learn their way around the sky. The counter-argument is that less time spent searching for a target object means more time observing it (and, of course, programmed supernova patrols are based on exactly this technology, so there are serious applications, too). My own preference is for the traditional star-hopping approach (p.39): there is considerable satisfaction to be derived from finding a "new" target for oneself. In the end, to GOTO or not to GOTO is a matter of personal choice.

## Finders

Under most circumstances, telescopic location of objects at high magnification is difficult. It is usual to start with a low-power field of view, increasing the magnification once the target has been identified and centered. Even at $\times 20$, though, the field of view in a telescope can be fairly restricted, so most observers have a small, very low-power secondary telescope mounted on the side of the main tube as an aid for locating objects. Typical finder telescopes (or "finderscopes") might be $6 \times 30$ or $7 \times 40$ short-tube monoculars; some observers may even use half of a pair of binoculars! Ideally, the finder will have visible cross-hairs or a target circle in the eyepiece to clearly show the aiming direction. Also popular as finders are non-magnifying Telrad-type devices, which use a mirror arrangement to project into the observer's field of view a red target which is seen against the sky.

Whatever your choice of finder, it is worth taking a few minutes to ensure that it is aligned as closely as possible with the main telescope's field of view. This can be done using a distant target (church steeples are a favorite) during daylight, or a bright star at the start of an observing session.

## Focal Length and Speed

The focal length of a telescope is the distance from the objective or primary mirror to the point at which light is brought to a focus. Of relevance to deep sky observing is the *focal ratio*, which is the focal length divided by the aperture. For example, an 80 mm lens which brings light to a focus at a distance of 800 mm – a fairly typical long-tube small refractor – this ratio is 10, by convention written as f/10. For an 80 mm lens with a focal length of 400 mm, the focal ratio is f/5. By analogy with photographic lenses, the f/5 instrument is said to be "faster," which is to say that objects appear brighter and show higher contrast than in the f/10. In my experience, the shorter-tube telescope is preferable for deep sky observing, though it won't readily deliver good high-magnification views of other targets such as the planets. The same argument applies to reflectors – many observers say that an f/6 Newtonian is their ideal deep sky instrument.

## How faint can you see?

A telescope's aperture determines its light-gathering power, and the more aperture, the fainter the objects that can be detected. The faintest that can be reached with a certain aperture is known as the *limiting magnitude*. Table 1 lists the limiting magnitudes for telescopes of different apertures under ideal conditions. Remember, though, that for extended objects the limit may be as much as a magnitude brighter than for stars, and also that atmospheric extinction (p.23), will further raise the magnitude limit for objects low down in the sky.

| TABLE 1: LIMITING MAGNITUDE FOR TELESCOPES | |
|---|---|
| Telescope Aperture (mm) | Limiting magnitude |
| 50 | +11.2 |
| 60 | +11.6 |
| 70 | +11.9 |
| 80 | +12.2 |
| 100 | +12.7 |
| 125 | +13.2 |
| 150 | +13.6 |

This book is concerned mainly with objects visible in 80–100 mm refractors, or reflectors up to 150 mm aperture; as the table indicates, this easily takes us down to deep sky targets of 10th magnitude, of which there are plenty.

## Eyepieces and Magnification

As important as the light-gathering objective or primary mirror in any telescope used for deep sky observing is the eyepiece used to magnify the focused image: the best telescope in the world will provide poor views if it is used with low-quality eyepieces. Manufacturers of quality eyepieces strive to eliminate the false-color fringing of chromatic aberration by combining individual elements of different glass types. Minimizing the air gaps between the elements helps to cut down on "ghost" images produced by internal reflections, while light transmission is improved by coating the optical surfaces with substances such as magnesium fluoride (visible as a

► *Most observers like to have several eyepieces to hand, offering a range of magnifications. Use of a Barlow lens (top center) can add to the versatility of an eyepiece collection. Filters, which screw into the eyepiece barrel (the author's O III filter is in the foreground), can enhance contrast and detail in some objects.*

"bloom" on the glass when viewed from the side). The lens nearest the eye is known as the eye lens; the lens at the opposite end of the eyepiece assembly is called the field lens.

A seemingly bewildering range of eyepiece types is available, and some modern designs have been created specifically for deep sky use. Some, like the multi-element *Nagler* type, are very expensive, but provide wonderful flat-field "picture window" views, and will perform superbly on telescopes with an aperture of 200 mm or more – beyond the range catered for in this book.

For general viewing, modern four-element *Plössl* eyepieces are more than acceptable, and are commonly supplied as part of the package with commercially produced telescopes. Such eyepieces provide good *eye relief* – in other words, the image is formed well beyond the eye lens, so the observer's eye does not have to be right up close to the glass; this is an advantage for wearers of spectacles in particular.

Most observers like to have a selection of eyepieces to hand, offering a range of magnifications. Different objects, and types of object, will each have their own response to magnification: galaxies may fade into almost invisible, featureless hazes in a high-power eyepiece, whereas the same eyepiece will resolve a globular cluster into a rich, clearly seen cloud of pinprick stars.

The magnification provided by an eyepiece can be found by dividing the focal length of the telescope by the focal length of the eyepiece: the latter is usually printed or engraved on the barrel of the eyepiece. For example, when used with a telescope of 600 mm focal length, a 20 mm focal length eyepiece will offer a magnification of ×30. A useful selection of eyepieces for deep sky observing might offer magnifications of ×20, ×60 and ×100, with views ranging from a wide field suitable for spread-out open clusters, to more highly magnified narrow fields allowing more detailed examination of the structure of globular clusters or planetary nebulae.

An accessory that allows the observer to effectively double their available range of magnifications is the *Barlow lens*. A "negative" optical element fitted into the focusing mount in front of the eyepiece, a Barlow lens extends the focal length, usually by a factor of two or sometimes three. A ×2 Barlow used with the eyepiece set described above allows additional magnifications of ×40, ×120 and ×200 to be obtained.

A point to bear in mind is that images are generally dimmer at higher magnifications. In particular, there is a limit to how much magnification a telescope can bear. A long-used rule of thumb is that the highest magnification at which a telescope can still give a viewable image is about ×50 per 25 mm of aperture. So, for an 80 mm refractor ×160 is the upper limit, while a 150 mm reflector should still provide decent images up to a magnification of ×300. Up to a point, the dimming effect at higher powers can work to the observer's advantage. Deep sky observers often find that higher powers make faint galaxies, for example, more visible by increasing the contrast: in many cases, the sky background appears darker relative to the target object at a high magnification.

Eyepiece fields of view can be described in a couple of ways. First, there is an eyepiece's *apparent field of view*, sometimes also called the acceptance angle. This can be visualized by holding the eyepiece some way from the eye, and looking at a blank piece of daylit sky – the angular extent of the circle of sky seen through the eyepiece is quite large, typically between 30° and 50° (for specialized eyepieces like the Nagler, the apparent field can be as much as 80°).

The actual field of view – the amount of sky the observer sees through the eyepiece at the normal viewing position – is much smaller than the apparent field. The actual field can be found by dividing the apparent field by the magnification – for example, the ×20 Plössl eyepiece on my 80 mm "spotter" telescope has an apparent field of 52°, giving an actual field of 2° 36'. It is easy to determine the actual field by timing how long it takes for a star close to the celestial equator to cross from one side of the view to the other with the telescope static (with the drive switched off). Delta Orionis (the westernmost star in Orion's Belt), Gamma Virginis and Theta Aquilae are useful for this purpose: the field diameter in arcminutes is the time in minutes taken for the star to complete its transit through the view, multiplied by 15.

### Filters

Most of the deep sky objects described in the chapters that follow can be seen perfectly well with conventional binocular or telescopic aid, without the use of special filters. For some targets, however, it can be an advantage to use a filter to improve contrast with the background sky, or to allow detection of light at a single wavelength which might be predominantly emitted by the object. As an example of the latter, planetary nebulae emit strongly at the 500.7 and 485.9 nm wavelengths of doubly ionized oxygen – the so-called O III lines. Filters that transmit only light at these wavelengths, and also reduce background light pollution (of which more later), can make some of the more elusive planetaries easier to see. A good *O III filter* will cost as much as a reasonable-quality eyepiece, but may be a worthwhile investment if planetary nebulae are your quarry of choice.

*Ultra-high-contrast* (UHC) *filters*, sometimes called *nebula filters*, pass the O III emission lines, together with hydrogen-alpha (656.3 nm) and hydrogen-beta (486.2 nm). While the dark-adapted eye isn't really sensitive to hydrogen-alpha, the greenish hydrogen-beta component is detectable, and can be a significant part of the light from emission nebulae (H II regions, Chapter 5). As an all-round filter for use with small telescopes when observing diffuse and planetary nebulae, many manufacturers recommend a UHC filter as preferable to the more wavelength-restricted O III type. The two are similar in cost.

*Light pollution reduction* (LPR) *filters* are popular with observers who have to contend with neighborhood streetlights. These filters cut down the transmission of troublesome background light from, for example, sodium lamps, but at the expense of slightly dimming the target objects.

In recent years, relatively inexpensive contrast-boosting filters have appeared on the market. These function mainly by ameliorating the effects of chromatic aberration in short-focus "fast" achromatic refractors, and are mainly of benefit for planetary observation.

Filters are threaded into the open, field lens end of the eyepiece assembly. They are readily available for the standard 32 mm (1.25-inch) eyepieces in widespread use. Users of 51 mm (2-inch) eyepieces (such as Naglers and other wide-field types) will have to purchase larger and therefore more expensive filters.

### Know Your Telescope

Observational astronomy, like any other pursuit, is something in which you will improve with practice. The better you know your observing equipment, the easier it will become to locate objects. Time devoted to getting used to the telescope's particular mount, and how it moves, is time well spent. Familiarity with the angular field provided by each of your eyepieces will make it easier to follow the instructions for locating objects given in the following chapters.

It is also important to be familiar with the orientation of the field. In astronomical telescopes, it is normal for the view to be inverted (telescopes for terrestrial use have an extra internal correcting lens to provide an erect view; this is also true of many "spotter" telescopes used for wide-field viewing as these are also sold in large numbers to, for example, bird-watchers). A normal astronomical telescope will thus give a field with north toward the bottom, and west to the left, when used by observers in the northern hemisphere (while those in the southern hemisphere will have south to the bottom of the field, and west to the right).

Use of a star diagonal in refractors or SCTs gives an erect image, but with east and west reversed: in the northern hemisphere, this gives a view with north to the top and west to the left (in the southern hemisphere, south at the top and west to the right). If you are in any doubt about which way is west, remember that, regardless of where on Earth you are, west is the side of the field of view at which stars exit the field.

# Observing

## Atlases and Catalogs

Star maps are essential for deep sky observing. A good, basic large-scale atlas showing stars to the naked-eye limit of magnitude +6 is a must for most amateur astronomical pursuits: *Norton's Star Atlas* is a long-established standard. The constellations and smaller-scale star patterns are a basic framework for object location. To find faint objects, however, it is usually necessary to use stars fainter than the naked-eye limit as a guide, and for the more serious pursuit of deep sky targets (or, for that matter, faint comets and asteroids) an atlas with a fainter – deeper – limit is desirable. Both *Sky Atlas 2000.0* (showing stars to magnitude +8.5) and *Uranometria 2000.0* (two volumes, to mag. +10) are useful.

Excellent star-plotting software is available for personal computers (there are *Redshift*, *TheSky* and *Starry Night*, among others; *Megastar* is popular with deep sky specialists). The observer can print off individual charts before a session, specifically for use on a particular target or range of targets, and so avoid the need to expose an expensive paper atlas to the cold and damp outdoors.

Throughout the descriptions of deep sky objects in this book, reference is frequently made to object designations of M (Messier's catalog), NGC (the *New General Catalogue*) and IC (the *Index Catalogue*). The eighteenth-century catalog drawn up by the French astronomer Charles Messier (1730–1817) contained 104 (extended by later astronomers to 109) objects in total; it was compiled as a listing of "fuzzy" objects which could be confused with his preferred prey of comets. Many of the Messier objects are bright, and it is worth bearing in mind that the telescopes originally used to compile this list were optically far inferior even to small modern amateur instruments.

The NGC (its full title is *New General Catalogue of Nebulae and Clusters of Stars*), published in 1887, was compiled by the Dane J.L.E. Dreyer (1852–1926) and is based heavily on previous work by the Herschels and other observers. Additions and corrections to the NGC were listed in the *Index Catalogue*, also by Dreyer, published in two parts in 1895 and 1907. The NGC contains 7840 entries, while the IC adds a further 5836; although some of these entries have been deleted as spurious, there is more than enough material to keep the serious deep sky enthusiast occupied for a lifetime. Printed and on-line versions of the NGC/IC are available, and a full Messier listing can be found at the back of this book.

## *Star-hopping*

A time-honored method for locating objects with an altazimuth-mounted telescope is the procedure of star-hopping. As the name suggests, it consists of starting at a bright landmark ("skymark?") star, then following patterns of fainter stars to arrive at the target. Having good charts to hand is a considerable boon for star-hopping; some are

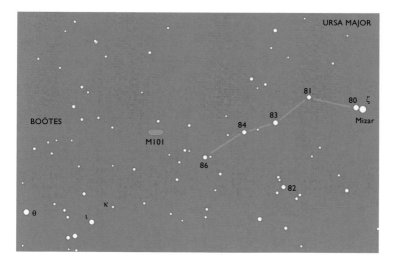

▲ A star-hop eastward across the sky from Mizar (the central star in the Big Dipper's handle) to the face-on spiral galaxy M101. The field shown here is 10° wide. It highlights the chain of stars 81, 83, 84 and 86 UMa, along which one can hop to arrive at the galaxy.

given in this book, while the detailed charts in *Sky Atlas 2000.0* and *Uranometria 2000.0* can be thoroughly recommended. Planning the star-hop in advance is important to success, and is best done well before the observing session.

An example of a relatively straightforward star-hop is to the 8th-magnitude face-on spiral galaxy M101 (NGC 5457) in Ursa Major, a little to the north, in the sky, of the Big Dipper's handle. Indeed, binocular observers will find that the handle and the galaxy (dim, but visible in 10 × 50s, say) can be comfortably accommodated in the same field of view. A binocular hop can start at Mizar in the handle's middle: 2nd-magnitude Mizar and its 4th-magnitude companion Alcor comprise one of the best-known double stars in the sky. Trailing away southeastward from Mizar and Alcor is a 5°-long chain of four 5th-magnitude stars. A 1.5° hop northeastward from the last star in this chain (designated 86 Ursae Majoris) takes the observer to M101. Careful examination should reveal it, on a good night, as a low-contrast diffuse patch with an apparent diameter only slightly less than that of the Moon. This is one of those objects where averted vision may help in detection.

The binocular star-hop is fairly straightforward, with all the stars laid out in the field of view. Doing the same thing with a telescope is a bit trickier, and will have to be done one star at a time. A low-power check in binoculars might be helpful before starting. When star-hopping with the telescope, start with a low-power eyepiece giving a 30- or 60-arcminute field of view; once the target (or at least its position) has been acquired, you can increase the power for a more detailed view.

This, of course, is why it is so useful to become familiar with the field-scale of your telescope: knowing that a shift in whatever direction of one field-width corresponds to a particular angular distance on the sky helps in relating the view to the scale of the finder chart. When using larger-scale charts, some observers employ a transparent overlay marked with a circle to indicate the angular size of the field of view in their favored eyepiece.

### Where and When to Observe

With a few notable exceptions, deep sky objects are faint and demand the best possible observing conditions to be seen to full advantage. Consequently, deep sky observers prefer to seek out the darkest, most transparent skies they can find.

*Light pollution* is a major problem for many observers in heavily populated areas. In suburban locations, direct streetlighting or light from neighboring houses can severely restrict observing opportunities. Even in the countryside beyond city limits, light cast skyward from ill-designed outdoor fittings (at filling stations, for example) generates an orange pall which dims the stars and makes faint deep sky objects – particularly diffuse nebulae – difficult, if not impossible, to detect. For die-hard enthusiasts, the only solution is to make observing trips to locations as far away as possible from urban skyglow; some North American observers head for the dry, dark deserts of the southwestern United States, for example. Even in the relatively crowded British Isles, though, determined observers can still find pockets of reasonably dark rural skies. A fair rule of thumb is that a site from which the Milky Way can easily be seen with the naked eye on a moonless night should be perfectly acceptable for general deep sky observing. If stars of fifth magnitude – such as some of the fainter ones making up the outline of Ursa Major – are visible to the naked eye, the site should be adequate.

Artificial light pollution is, alas, a semipermanent problem. The sterling efforts of organizations such as the International Dark Sky Association and the UK's Campaign for Dark Skies have increased awareness of the problem and brought about some significant local improvements.

As important as artificial light pollution is, of course, the effect of moonlight. A bright Moon will lighten the sky background and swamp faint objects. Double stars can still be observed under such conditions, but for most deep sky enthusiasts the ten-day interval centered on full Moon each month is pretty much enforced down-time.

Climate is also an important factor, especially when coupled with moonlight or artificial light pollution issues. Hazy skies often prevail in North America and Northwest Europe during the autumn months, and at other times when calm, relatively stationary high-pressure systems (anticyclones) dominate the weather. A hazy sky, full of small suspended water droplets, is, hardly surprisingly, less transparent than a clear, dry sky, and will hinder the detection of faint objects. This is one reason why some observers prefer desert skies. Additionally, suspended haze scatters light pollution and can worsen its effects markedly.

Another important factor for observers at higher temperate latitudes, such as those of the British Isles, is the seasonal summer twilight. During June and July, when the deep sky riches of Scorpius and Sagittarius are best presented, the sky in the UK north of the Midlands never becomes fully dark. The Sun at this time is never more than 12° below the horizon for observers at such latitudes (above 52°N), and twilight persists throughout the short night. The problem is particularly acute for observers in Scotland (and worse still in Scandinavia).

In maritime climates like that of the British Isles, clouds and moisture are never usually far away. Regular observers in such places learn to keep an eye on the weather forecast, and to make the most of good opportunities as and when they arise. Excellent clear skies can follow daytime showers, for example, while the best observing conditions often follow immediately after the passage of a cold front. These conditions offer sparkling views of the stars and faint objects, but can have a downside. Planetary observers will attest to the turbulence of the air on such nights – conditions described as poor *seeing* – which can turn the image of Mars or Jupiter into a shimmering blob in the eyepiece, and can also have a negative effect on the visibility of some types of deep sky object, for example the disks of planetary nebulae.

The ideal deep sky night, then, will find the observer at a dark location at a time when the Moon is well out of the way, preferably during a spell of dry weather with little or no haze. For most observers such a combination of conditions is rare, and perhaps the most useful advice is to make the best use of whatever clear dark skies you can find, whenever you can.

### Planning an Observing Session

On any clear, dark night, the deep sky observer will have hundreds of targets from which to choose. It is a good idea to plan well beforehand which objects to observe, rather than randomly "potting." Ahead of the session, draw up a list of maybe ten objects to be covered on the night. Some targets will be easier to find than others, and expect to spend 10 to 15 minutes tracking down some of the fainter ones. Taking time to examine, make notes about and sketch the field of a deep sky object once it has been found might give a perfectly sensible "haul" of four or five objects per hour. It is better to make quality observations of a few objects rather than tearing through a large selection, none of which is particularly closely studied – astronomy isn't a competitive sport, and there are no prizes for bagging more objects than anyone else.

The target list for the night will depend on the season, of course. For instance, May nights are good for viewing globular clusters in the Ophiuchus/Scorpius region, while open clusters in the Auriga to Monoceros stretch of the Milky Way are ideally viewed in December and January.

To see objects at their best, choose a time when they are reasonably close to the meridian. This will often lead you to draw up a list of objects in order of increasing RA, as the east–west motion of the sky will carry them to the meridian in turn. For example, a September evening's observing might begin with the globular clusters in Capricornus, Pegasus and Aquarius, followed by some of Aquarius' fainter objects, winding up, a good couple of hours later, with the Andromeda Galaxy. Many observers like to start with an easy object (quite often an already-observed favorite), seek out the trickier targets in midsession and end with another easy or favorite object. Not every attempt to find an object will necessarily be successful (at least on the first occasion): patience is very much the deep sky observer's by-word.

Additional preparation will be required if you are traveling to a remote site in search of dark skies. For safety reasons, a first aid kit and a mobile telephone are recommended items in case of emergency. Before setting off, it is a good idea to have a checklist of things to take – flashlight, batteries, notebook, atlases, telescopes/binoculars (it *has* been known for observers to forget these), pens/pencils, and so on.

## Keeping a Record

Once the target has been located, take some time to study it. According to one school of thought, you will not have properly observed an object until, at the very least, you have made a few notes and visual impressions for future reference. Better still, perhaps, make a field sketch. A permanent record of the object as seen in your binoculars or telescope, kept in your own observing log, lends a sense of "ownership." There is certainly a world of difference between "just looking" at an object and making a proper, thorough observation which will leave a lasting impression of its appearance.

I recommend keeping an observing log, whatever your astronomical interest, as a diary of nights spent under the stars. It can be interesting – and informative – to review past observations of some objects, comparing their appearance in different telescopes for example. Also, you should find that as you become more experienced you will be able to tease more detail from certain objects than was apparent on an earlier visit: it is worth returning to deep sky objects at a later date for another look.

An observing log is also a good place to keep frequently needed information – the latitude and longitude of your observing site(s), for example, and details of telescopes and eyepieces. The inside cover of

▶ *Basic sketching essentials – a clipboard and paper, plenty of pencils (some blunted for shading), an eraser and cotton buds for "smudging." Use of the red flashlight conserves night vision.*

the log book is an excellent place to list the fields of view offered by your eyepieces for quick reference.

### Sketching

The views through typical amateur telescopes differ greatly from images recorded on film and obtained with CCD cameras with large professional instruments. In many cases, the best means of recording deep sky objects is to make a simple pencil sketch, showing what you can see (importantly, not what you *expect* to see: don't prejudge an object's appearance from images in glossy magazines and coffee-table books). It is convenient to use a circle to represent the field of view, centered on the target object. A 50 to 55 mm circle is ideal for this purpose, and it is easy to prepare several of these observing blanks in advance, with a pair of compasses. I usually start by drawing in the brightest field stars and their relative positions. Deep sky sketchers work in "negative" – stars are drawn as black dots. Fainter stars are then added, and relative brightness is indicated by using smaller dots for fainter stars. With a little practice, the observer soon becomes quite proficient at reproducing stars' relative positions by spotting triangles and other geometrical patterns.

How the object itself is drawn will depend on its nature. A loose, well-resolved open star cluster will be represented simply as a collection of dots indicating the positions and brightnesses of the constituent stars. Globular clusters such as M13 appear as concentrated fuzzy balls at low power, becoming partly resolved into their component stars in large telescopes at higher powers. Drawing the relative positions of thousands of individual stars in a globular cluster is not possible, but a general impression of more concentrated radial "arms" or condensed regions can be given. At low powers, some planetary nebulae will appear as little more than fuzzy spots, surrounded by the sharper points of the field stars.

Various artistic tricks are employed by skilled observers to capture the appearance of deep sky objects. A medium-hardness HB pencil with a sharp point is good for marking star positions. A blunt HB or soft B pencil can be used to draw diffuse nebulae, whose appearance can be further softened by finger smudging, or by using a blurring tool called an artist's stub, available from art suppliers; small wads of paper or even cotton buds (Q-Tips) can be used to obtain the same effect.

It may take several attempts to get the sketch just right! Don't be afraid to scrap one which is going wrong and start again, or to make liberal use of an eraser to correct minor errors. Make notes on the draft sketch, too, for later guidance. The sketch made at the telescope should be the basis for a neater, final version worked up indoors in more con-

▶ *The author's sketch of the Leo galaxy NGC 2903 (p.60), viewed through his 80 mm f/5 refractor at ×40. The galaxy appeared diffuse with a noticeable concentration south of center, and was elongated in PA 020–200°. From the sketch and the known 78′ field, the galaxy has an estimated visual long dimension of 8′.*

venient conditions for artistic endeavor, preferably soon after the original observation when the details are still fresh in the mind.

Keep a note of the date and time of the observation, and the instrument and magnification used, and also ensure that the orientation of the field is clear by marking due north or due south (bearing in mind that this will not always be at the exact top or bottom of the field of view – it relates instead to the direction from the field to the appropriate celestial pole). Written notes to accompany the sketch in your observing log can include descriptions of features such as the apparent elongation of an object, and in which direction, the presence in a globular cluster of a dense or diffuse core, hints of nebulosity, resolution of tightly packed stars, and so forth. Some deep sky objects have interesting star patterns or strongly colored stars in their fields of view; these are features worth noting.

From the known size of the telescope or binocular field, it might often be possible to estimate the angular diameter of the object viewed: for example, a globular cluster spanning one-tenth of a 30-arcminute field can be taken to have an angular diameter of 3 arcminutes. Alternatively, this can be done after the observation by comparison with the angular scale of a detailed star atlas. Objects such as faint galaxies will typically appear rather smaller than their catalog diameters when viewed in small telescopes: catalog values are often based on images to very faint magnitude limits taken with large professional instruments, and include faint peripheral regions.

# 3 · GALAXIES

To cosmologists studying the origin and evolution of the Universe, galaxies are fundamental units. Galaxies formed early in the history of the Universe, appearing within 200 million years of the Big Bang moment of creation, which is now estimated to have occurred 13.7 billion years ago. Clouds of hydrogen and helium, the lightest chemical elements and produced in the Big Bang, condensed to form the first generation of stars, groups of which are believed to have aggregated into structures similar to (or even identical with) globular clusters, which in turn coalesced into galaxies. Nuclear fusion in massive stars produced elements heavier than hydrogen and helium. These heavy elements were then released by supernova explosions into the interstellar medium early in the galaxies' history, in turn gathering into the gas and dust clouds from which later generations of stars formed, and continue to form today. The closest galaxies accessible to our amateur telescopes are ancient, evolved units. Evidence for vigorous star formation in galaxies in the early Universe is found in the famous Hubble Deep Fields of 1995 and 1998, images on which galaxies were recorded to a lookback time of 12 billion years.

▼ *The Hubble Ultra Deep Field, published in March 2004, shows galaxies in the very early Universe, 400 to 800 million years after the Big Bang. This is an even greater lookback time than that of the Hubble Deep Fields (p.11).*

▶ A typical spiral galaxy has a large nuclear bulge containing old (Population II) stars, surrounded by wound-up spiral arms in a flat disk where new stars (Population I) form. Surrounding the disk and bulge is a halo of Population II stars, including concentrated aggregations of stars in globular clusters.

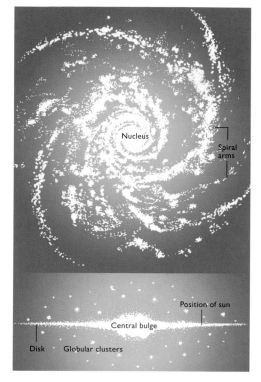

Galaxies show a range of morphologies. Some are described as *irregular*, having no obvious organized structure. Others, like M87 (NGC 4486) in Virgo, are vast, usually somewhat flattened, balls of hundreds of billions of stars described as *giant elliptical* galaxies, showing no internal structure. The well-known Andromeda Galaxy (M31, NGC 224) is a classic example of a *spiral* galaxy, with several spiral arms rich in gas, dust and stars in a flat disk wrapped around a central, nuclear bulge. Many spiral galaxies show a pronounced bar structure running across the central bulge and into the arms. Our own Milky Way is thought to be such a *barred spiral*.

The arms of spiral galaxies are believed to mark density waves propagating around the galaxies as they rotate. As they sweep up and compress material, density waves trigger new rounds of star formation in the spiral arms. Studies during the 1940s led the German-American astronomer Walter Baade (1893–1960) to identify two principal stellar *populations* in the Milky Way. Population I stars are relatively young, and are found in the spiral arms or the disk, while the more ancient Population II stars reside in the nuclear bulge and the less dense halo

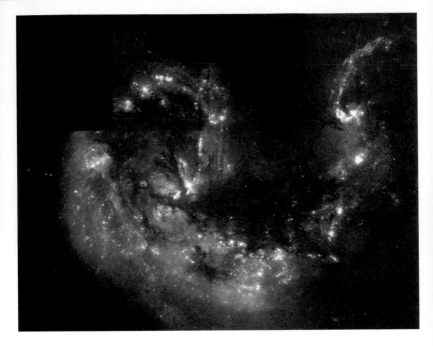

▲ The interacting galaxy pair of the
Antennae (NGC 4038 and 4039) in
Corvus, where tidal forces have triggered
vigorous star formation.

that surrounds the disk. The same stellar population distribution is found in other spiral galaxies. Giant elliptical galaxies contain very little dust and have only Population II stars – new star formation does not go on in such galaxies. Irregular galaxies may be the end-product of interactions between two galaxies that experienced a close encounter. Such encounters can tidally disrupt galaxies (as seen, for example, in the interacting pair known as the Antennae, NGC 4038 and 4039 in Corvus), and set off vigorous waves of star formation.

Spiral galaxies are classified according to how tightly wound their spiral arms appear. They range from tight spirals with a large nuclear bulge (Sa) to loose spirals with a relatively small nuclear concentration (Sc). An equivalent classification is used for barred spirals, from SBa to SBc. Elliptical galaxies are classified from E0 (spherical) to E7 (highly flattened). In 1936, the American astronomer Edwin Hubble (1889–1953) proposed that there was a developmental progression from E0 galaxies to flattened disks lacking spiral arms (S0, *lenticular* galaxies), from which increasingly loosely wound spirals or barred spirals would form, in turn followed by irregular galaxies. While useful as a basic classification system, Hubble's "tuning fork" diagram of galactic form is no longer considered to represent an evolutionary sequence. Irregular galaxies are denoted by "Irr."

Galaxies also show a range of sizes and star content. The Andromeda Galaxy is a fairly typical Sb spiral, containing 300 billion stars in a disk 150,000 light years in diameter. Others are much smaller: dwarf galaxies containing only a few hundred thousand stars may be very numerous in the Universe, but only those fairly near our own Galaxy can presently be detected, even with large professional telescopes. There is strong evidence, from the presence of "star streams" in the Milky Way, that our Galaxy is in the process of consuming material from dwarf galaxies in its gravitational sphere. The giant ellipticals are thought to be products of yet more voracious galactic cannibalism.

Some larger galaxies, such as our Milky Way and M31 in Andromeda, are orbited by small satellite galaxies. The Large and Small Magellanic Clouds are the largest satellites of our Galaxy. Other dwarf galaxies associated with our own appear to be undergoing tidal disruption, their constituent stars being pulled from them to join the Milky Way: examples include the Sagittarius Dwarf Elliptical Galaxy and the Canis Major Dwarf Galaxy. M31 has four dwarf elliptical satellites.

At the centers of the largest galaxies, where the density of stars, gas and dust is very high, it is likely that supermassive black holes reside. Studies of stellar motions in the heart of the Milky Way provide compelling evidence for the presence of a black hole at the core of our home Galaxy.

The deep sky objects discussed in subsequent chapters are representative of features present in other galaxies, but are much closer to home. The central bulge of our Galaxy is evident on June and July

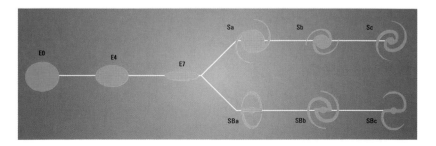

▲ Edwin Hubble's "tuning fork" classification of galaxies is no longer believed to represent an evolutionary sequence. It is still useful, however, as a broad system for describing the morphology of galaxies.

evenings in the constellation Sagittarius, with many rich starclouds and open clusters on view. A halo of globular clusters, similar to that which surrounds the hub of our Galaxy, is also found around M31. Star-forming H II regions (p.95), rich in nebulosity, can be seen in the spiral arms of other galaxies – the counterparts of objects such as the Orion Nebula. Dark lanes of obscuring dust analogous to that which splits the Milky Way in Cygnus/Aquila are spectacularly evident in some edge-on galaxies, such as NGC 4565 in Coma Berenices. Detailed images of M31 show clouds of stars which have recently (on the cosmological timescale) formed in its spiral arms.

In the descriptions that follow, galaxies are often discussed in groups. This is more than a coincidence: galaxies tend to be found in gravitationally connected clusters. Our Galaxy, for example, is a member of a cluster of more than 30 members known to astronomers as the Local Group, which also includes the Andromeda Galaxy and M33 in Triangulum. A major grouping of galaxies is found in the direction of the constellations Virgo and Coma Berenices; the giant elliptical M87 lies near its heart, and many other galaxy clusters appear to have such an object at their center.

Galaxies can be frustrating objects for the visual observer. Don't expect to see any of the exquisite structural detail or color often apparent in long-exposure photographs or CCD images, even in the eyepiece of a very large telescope. As a prime example, visual observers should have no difficulty picking out the central bulge of M31 in binoculars or a small telescope, but the spiral arms – which have a lower surface brightness – are barely visible to the eye, even in large-aperture instruments. It is, however, still very worthwhile looking into the eyepiece and establishing just what you *can* see.

◄ A deep image of part of the rich galaxy cluster known as the Coma Cluster. It shows a couple of prominent giant elliptical galaxies.

Of all the deep sky objects, galaxies (together with diffuse nebulae) are the most demanding of dark, transparent observing conditions. The requirement is most critical for face-on spiral galaxies, which appear as extended circular or oval hazes of low surface brightness; those whose disks are more tilted toward us are easier to detect.

Drawings of galaxies and their field stars are worth making, allowing the observer to estimate apparent diameter, orientation and outline. Some galaxies show a star-like nucleus, while others have a more even surface brightness: these structural details should be recorded. Observers using larger-aperture instruments may detect dark lanes or mottling (the latter indicative of H II regions and starclouds in spiral arms) in some of the brighter galaxies under good conditions.

In general, larger apertures will reveal more of a galaxy's outer regions. It is an interesting exercise to compare diameter and appearance in a range of instruments, and this is part of the appeal for many enthusiasts of star parties and observing sessions organized by local astronomical clubs. As mentioned in Chapter 2, sometimes moving up to a higher-magnification eyepiece can improve the view of a galaxy. For instance, I find that in my 80 mm short-focus refractor the "Leo Trio" of M65, M66 and NGC 3628 look rather more imposing and contrasty in the darker field at ×40 than in a ×20 view. It is certainly interesting to compare the view of each object at a range of magnifications.

Messier's catalog of "nebulous" objects includes 40 galaxies, and a great many more listed in the NGC and IC are accessible with amateur astronomers' instruments. The selection in this chapter includes some of the brighter examples, and an observer equipped with a very large telescope will find hundreds more within easy reach.

## Selected Galaxies

| Large Magellanic Cloud | | | | |
|---|---|---|---|---|
| Dorado | RA 05h 23.6m | dec −69° 45′ | mag. +0.1 | Map 8 |
| Small Magellanic Cloud | | | | |
| Tucana | RA 00h 52.7m | dec −72° 50′ | mag. +2.3 | Map 8 |

Easily visible to the naked eye as diffuse patches resembling detached pieces of the Milky Way, the Magellanic Clouds are named for the Portuguese explorer Ferdinand Magellan (1480–1521), during whose circumnavigational voyage of 1519 they were first described by European observers (though they had, of course, been known to inhabitants of the southern hemisphere since time immemorial). The Magellanic Clouds are satellites of our Galaxy; the Large Magellanic Cloud (LMC) is at a distance of 179,000 light years, while the Small

▲ *The Large Magellanic Cloud (LMC) is a satellite of our Galaxy, and a prominent naked eye object for observers at southerly latitudes. It shows* *hints of a disrupted bar structure. At left (the eastern end of the LMC ), the Tarantula Nebula (NGC 2070) is visible.*

Magellanic Cloud (SMC) is 210,000 light years away. Neither is visible north of 20°N latitude, and they are certainly best seen from the southern hemisphere.

The LMC is a dwarf irregular galaxy, although in wide-field photographs it shows features suggestive of a disrupted barred spiral. Located mostly in the constellation Dorado, with its center roughly 12° SW of the mag. −0.6 star Canopus, the LMC contains perhaps 10 billion stars. Small telescopes reveal a wealth of detail in the form of open star clusters and nebulosity. The LMC's most noted object is the bright Tarantula Nebula (NGC 2070, p.103) close to its eastern side, and near where Supernova 1987A exploded. The LMC's angular diameter is about 11° × 9°, giving a true size of around 25,000–30,000 light years. Tidal disruption by our Galaxy has distorted the LMC, and a close brush between the two about 200 million years ago triggered a vigorous episode of star formation in the LMC.

Located east of the LMC in Tucana, centered around 5° NNE of the mag. +2.8 star β Hyi, the Small Magellanic Cloud (SMC) is also classed as an irregular galaxy, having apparently been tidally disrupted by the gravitational fields of both the Milky Way and the LMC. Like its larger counterpart, the SMC contains numerous star clusters and some nebulosity, and is a rewarding object for small telescopes. Slightly to the south of the SMC, the globular cluster 47 Tuc (NGC 104, p.77) is a foreground object. The SMC covers an area of sky about 3° across, has an actual size of around 15,000 light years and contains a billion stars.

| M31 NGC 224 Andromeda Galaxy | | | | |
| --- | --- | --- | --- | --- |
| Andromeda | RA 00h 42.7m | dec +41° 16′ | mag. +3.4 | Map 7 |
| **M32 NGC 221** | | | | |
| Andromeda | RA 00h 42.7m | dec +40° 52′ | mag. +8.1 | Map 7 |
| **M110 NGC 205** | | | | |
| Andromeda | RA 00h 40.4m | dec +41° 41′ | mag. +8.1 | Map 7 |

The Andromeda Galaxy (M31), at a distance of 2.4 million light years, is widely considered to be the remotest object visible to the naked eye. It is easy to find by a naked-eye star-hop starting at 2nd-magnitude α And (Alpheratz) at the NE corner of the Square of Pegasus. From here, two lines of moderately bright stars trail eastward: following the more northerly through π to μ And – the second star to Alpheratz's east – then north by a couple of degrees to mag. +4.1 ν And takes the

▶ The central region of the Andromeda Galaxy M31, showing the prominent nuclear bulge. The companion galaxy M32 is at the right; M110 is at bottom left in this image.

observer to M31. Unless conditions are badly light-polluted or very misty, the Andromeda Galaxy stands out as a hazy 3rd-magnitude patch about 1.5° (three Moon-widths) west of $\nu$. This famous object has been known since antiquity: it is described in the catalog drawn up by the Persian astronomer al-Sufi (AD 903–986) as a "little cloud."

The Andromeda Galaxy is a familiar photographic subject, but its visual appearance in binoculars or a small telescope is limited mainly to the bright nuclear bulge: the exquisite, delicate spiral arms are revealed only by long-exposure imaging. M31 is an Sb galaxy; its large nucleus is surrounded by quite tightly wound spiral arms. It is inclined 13° from edge-on to us and has its long axis aligned NNW–SSE.

In the naked-eye view, M31's nuclear bulge is obviously extended. Estimates of its angular size in $10 \times 50$ binoculars are typically about half a degree. Under good conditions, binoculars will show the satellite galaxies M32 (24′ due south of M31's center) and M110 (a degree to the NW). They appear separated from M31's bulge: their apparent proximity to its edge in long-exposure images can mislead visual observers into looking for the satellite galaxies closer in.

The satellite galaxies are easy to see in small telescopes. Each would be regarded as a fine object in its own right even if they were not associated with M31. M32 is the smaller, with apparent angular size $11′ \times 7′$, corresponding to a true diameter of 6000 light years. M110 has angular dimensions of $20′ \times 13′$ and a diameter of 12,000 light years. M32 and M110 are, respectively, E2 and E5 elliptical galaxies, and their satellite status makes them analogous to our own Galaxy's Magellanic Clouds. Like the Milky Way, M31 has a surrounding halo of globular clusters, but only the largest amateur telescopes will offer much chance of seeing these.

M31 has a couple of more distant satellite galaxies, located in the neighboring constellation of Cassiopeia (NGC 147, RA 00h 33.2m, dec +48° 30′, mag +9.5, E4; NGC 185, RA 00h 39.0m, dec +48° 20′, mag. +9.2, E3).

The main effect of using binoculars or a small telescope is to increase the apparent size of the nuclear bulge to as much as 3° – six Moon-widths! – and making the satellite galaxies more visible. Larger apertures begin to reveal the fainter spiral arms as a haze surrounding the nucleus. In instruments of 100 mm aperture and upward, the star-cloud designated NGC 206 may be evident as a nebulous condensation in the spiral arm to the SW of the nuclear bulge.

In August 1885 a supernova erupted in M31, reaching a peak magnitude of +6. Given the variable star designation S Andromedae, it has the distinction of being the first such object ever seen beyond the confines of our own Galaxy.

M31 is larger than the Milky Way, with an overall diameter of 150,000 light years. Spectroscopic observations indicate that M31 is approaching the Milky Way: the two will collide – possibly merging to form a giant galaxy – in about 3 billion years' time. The detection of a double nucleus in M31 during Hubble Space Telescope observations in 1993 suggests that, relatively recently in cosmological terms, the Andromeda spiral has undergone a merger, probably with a dwarf companion galaxy.

## M33 NGC 598

Triangulum   RA 01h 33.9m dec +30° 39'   mag. +5.7   Map 7

The third-largest member of the Local Group of galaxies (after the Andromeda Galaxy and the Milky Way), M33 can sometimes be a difficult object despite its bright catalog integrated magnitude of +5.7. The galaxy is 4° west of mag. +3.4 α Tri, the sharp tip of the westward-pointing Triangulum. A very loose spiral (classed as Sc) with a small nucleus and lacking a marked central hub, M33 is presented face-on to us and its light is spread over an area of 67' × 42' – more than twice the apparent diameter of the Moon along the longer axis, which is aligned NNE–SSW. The loosely wound spiral arms have low surface brightness and thus poor contrast with the sky: on hazy nights in October or November, when M33 is at its highest for northern observers, it may prove impossible to see this object.

While it is fairly easy and obvious in binoculars, M33 can prove disappointing telescopically. In a 110 mm reflector at ×31 it is dim and diffuse with a slightly brighter central patch which averted vision shows at its best. The small central condensation has led some observers to nickname this the Pinwheel Galaxy. In a 150 mm refractor, I have seen M33 as a good, reasonably contrasty object at ×40, but on doubling the magnification it faded almost into the background. Since the lower magnification of binoculars spreads out M33's light rather less, they are often a better choice than a small telescope for viewing it.

M33 is 2.4 million light years away and has a diameter of about 50,000 light years. It was discovered in August 1764 by Messier. Lord Rosse (1800–1867), observing with his giant 72-inch (1.8-meter) "Leviathan" reflector at Birr in Ireland, was the first to note spiral structure in this object, during the 1840s. Denser concentrations along the spiral arms mark the presence of H II regions and star-clouds, some of which have their own NGC entries (NGC 604, RA 01h 34.5m, dec +30° 48', in the NW of the galaxy, is the most pronounced). A really large telescope (250 mm aperture or greater) is required to bring these out.

**M81 NGC 3031**

Ursa Major   RA 09h 55.6m dec +69° 04'   mag. +6.9   Map 1

**M82 NGC 3034**

Ursa Major   RA 09h 55.8m dec +69° 41'   mag. +8.4   Map 1

The finest galaxy pairing in the northern sky, M81 and M82 are quite easy to find, 5° east of the 5th-magnitude stars σ¹, σ² and ρ UMa, marking the Great Bear's "ears" in the popular portrayal where its snout points westward, to the north of the Big Dipper's bowl. The nearest naked-eye star, again of 5th magnitude, is 24 UMa, a couple of degrees to the galaxies' west. On a good, transparent night, M81 and M82 are readily visible in 10 × 50 binoculars: M81 is markedly the brighter, with a catalog magnitude of +6.9.

M81 and M82 lie 38' apart – easily contained in the same telescopic field at low magnifications (×40, say) – with M82 to the NE of its larger, brighter partner. M81 is an Sb spiral, somewhat tilted in its presentation. Binoculars show it as a hazy spot, while a small telescope will show the extended nuclear region with a "stellar" central condensation. In long-exposure images, M81 has dimensions of 24' × 13', with the longer axis oriented almost N–S; visually, in a medium–aperture amateur telescope, its apparent diameter is more like 10'. M81 has a true diameter of 70,000 light years.

A supernova was discovered in M81 at the end of March 1993. Designated SN 1993J, this Type II supernova (p.160) reached a peak magnitude of +10.5, bright enough to be visible in small telescopes. Most supernovae, in more remote galaxies, require fairly large apertures for detection.

M82 is an irregular galaxy, somewhat smaller than M81 and presented close to edge-on, appearing as an elongated bar with photographic dimensions of 12' × 6', oriented ENE–WSW; visually, the long axis is about 8'. At mag. +8.4, M82 is a faint binocular target, but it is readily visible in 60 to 80 mm aperture telescopes. A 9th-magnitude field star lies just south of M82's western tip. Deep images show

◄ The M81/M82 galaxy pair in Ursa Major, imaged by Nick Hewitt; M82 is uppermost in this view.

M82 to be dusty and disrupted: a cosmologically recent encounter with M81 has led to a wave of star formation, and this galaxy is a strong radio source. M82 has an actual diameter of 35,000 light years. The separation between M81 and M82 is only 150,000 light years.

M81 and M82 were discovered by the German astronomer Johann Bode (1747–1826) on December 31, 1774, and Messier added them to his catalog in 1781. They are part of a group of maybe 20 galaxies, which, at 12 million light years from us, is probably the closest cluster of galaxies beyond the Local Group. Another member is NGC 2403.

## NGC 2403
Camelopardalis   RA 07h 36.9m dec +65° 36'   mag. +8.5   Map 1

NGC 2403 was discovered by William Herschel (1738–1822) in 1788. It is located in a rather barren part of the sky in the dim northern constellation Camelopardalis, about 6° NW of o UMa, the mag. +3.4 "snout" of the Great Bear. The most useful field stars for location are the 6th-magnitude 49 and 51 Cam, separated by about 3°, to the galaxy's east: NGC 2403 is a couple of degrees west of 51 Cam, which makes the right angle of a triangle with the galaxy and 49 Cam. In July 2004, an 11th-magnitude supernova (the brightest since SN 1993J in M81) was discovered in NGC 2403's western spiral arm.

NGC 2403 is a face-on Sc spiral galaxy of low surface brightness, with open spiral arms and a bright nucleus. Its overall photographic dimensions are 18' × 11', but only the brighter central parts are visible in small telescopes. NGC 2403 is 10 million light years distant.

## IC 342
Camelopardalis   RA 03h 46.8m dec +68° 06'   mag. +8.4   Map 1

In addition to NGC 2403, Camelopardalis is home to the faint, face-on Sc spiral galaxy IC 342, 10 million light years distant. IC 342 is located 3° south of the wide pair of 5th-magnitude stars γ Cam and Hip 17854. A good guide for finding these is the eastern arm of Cassiopeia's W: extending the line between δ and ε Cas eastward by three times the distance between them brings the view to γ Cam.

Like NGC 2403, IC 342 has low surface brightness, and will require an instrument of at least 80 mm aperture to be well seen. The nucleus is prominent at the center of low-contrast spiral arms which give an overall circular diameter of about 20'; only the nucleus is prominent in small instruments. IC 342 is close to the plane of our Galaxy's disk, and is consequently dimmed by interstellar dust along the line of sight. Remarkably, the object eluded William Herschel's thorough celestial survey; it was discovered by William Denning (1848–1931, a noted English meteor, comet and planetary observer) in 1895.

## NGC 5128
Centaurus   RA 13h 25.5m dec −43° 01'   mag. +6.7   Map 4

Well known from frequent photographic appearances in books and magazines, the peculiar galaxy NGC 5128 is invisible to observers in Northwest Europe, and even from the southern United States it only just rises above the horizon. The best views are to be had from the southern hemisphere, where NGC 5128 is a relatively easy binocular object. NGC 5128 is famous for the prominent dark lane of material lying E–W across its middle, splitting the brighter core of the galaxy into two halves. This band can be detected in telescopes of 100 mm aperture.

The circular, 18'-diameter bright bulge of NGC 5128 is characteristic of a large elliptical galaxy, but the dark lane is more typical of a spiral galaxy's disk. Astronomers studying this object have proposed that the dark lane is, indeed, a spiral galaxy, in the process of merging with a brighter elliptical galaxy. The strong radio emission from NGC 5128 is consistent with such a nature – to radio astronomers, this object is known as the source Centaurus A.

NGC 5128's dark lane was the site of a supernova, discovered during visual searching by Rev. Robert Evans in May 1986. This Type I object (p.160), cataloged as SN 1986G, reached peak magnitude +12.5.

NGC 5128 is found 5° to the north of the globular cluster ω Cen (p.76), or 5° ESE of the triangle of 3rd-magnitude stars φ, ν and μ Cen. A fascinating galaxy, it was discovered by the Scottish astronomer James Dunlop (1795–1848) from the Brisbane Observatory in New South Wales, Australia, in August 1826, and lies 15 million light years away.

## The Leo Trio

## M65 NGC 3623
Leo   RA 11h 18.9m dec +13° 05'   mag. +9.3   Map 3

## M66 NGC 3627
Leo   RA 11h 20.2m dec +12° 59'   mag. +8.9   Map 3

## NGC 3628
Leo   RA 11h 20.3m dec +13° 36'   mag. +9.5   Map 3

Leo is home to a couple of well-known galaxy groupings, the brighter of which is the "Leo Trio" – M65, M66 and NGC 3628, located 3° SSE of the 3rd-magnitude star θ Leo, at the SW corner of the triangle of stars making up the Lion's hindquarters. A good guide for locating this group is a N–S line of three binocular-bright stars, the most northerly of which is the 5th-magnitude 73 Leo, the galaxy group lies just to the east. M66 can be seen quite easily in 10 × 50 binoculars on a reasonable night, with a faint foreground star visible against its hazy periphery. M65, to the west, is a bit trickier and usually requires averted vision.

This galaxy group is easier in a small telescope. My 80 mm wide-field refractor picks up all three without difficulty at ×20. As in the binocular view, M66 is the most obvious, with the foreground star visible in binoculars in its western edge. Boosting the magnification to ×40 separates galaxy and star, and makes M65, 20′ to the west, more obvious. All three galaxies are contained in the 1.4° field, with 73 Leo just out of the field to the west. The low-power view is dominated by a 6th-magnitude star 12′ to the north of M65.

At ×80, M66 appears fairly circular, while the fainter M65 is markedly elongated NE–SW. NGC 3628, 35′ NNE from M65, is elongated E–W and has quite a high surface brightness – it is perhaps the easiest edge-on galaxy for small telescopes. M65 has catalog dimensions of 6′ × 2′, and M66 5′ × 2′, corresponding to sizes of 80,000 and 75,000 light years, respectively. NGC 3628 is about 9′ × 1′ in angular size.

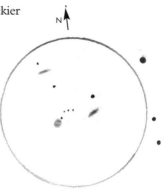

▲ A sketch by the author of the "Leo Trio" of galaxies, as viewed in an 80 mm f/5 refractor at ×40; the useful locating star 73 Leonis lies just out of field to the west. M65 lies west of M66, with the elongated NGC 3628 to the north of this pair.

This grouping is genuine, lying 30 million light years away. Deep images show the disk and dark lane of NGC 3628 to be warped by the gravitational influence of its neighbors. M65 is an Sa spiral galaxy, while M66, with its more prominent nuclear bulge, is a type Sb. NGC 3628 is an Sc spiral, less tightly wound. M65 and M66 were discovered by Messier's collaborator Pierre Méchain (1744–1805) in 1780.

| M95 NGC 3351 | | | |
|---|---|---|---|
| Leo   RA 10h 44.0m dec +11° 42′   mag. +9.7   Map 3 | | | |
| **M96 NGC 3368** | | | |
| Leo   RA 10h 46.8m dec +10° 49′   mag. +9.2   Map 3 | | | |
| **M105 NGC 3379** | | | |
| Leo   RA 10h 47.8m dec +12° 35′   mag. +9.3   Map 3 | | | |

Leo's second group of relatively bright galaxies lies about 9° east of Regulus, in a rather empty region of the sky. The best way to locate them is to star-hop eastward from 4th-magnitude ρ Leo (5° ESE of Regulus) to a tight equatorial triangle of stars (30′ to a side) with 5th-magnitude 53 Leo at its east. The field of M96 and M95 is just over a degree NNW of 53 Leo.

I find M95 and M96 beyond the reach of $10 \times 50$ binoculars, but they show well in an 80 mm refractor at $\times 20$. M96, the more easterly, is the brighter – obviously non-stellar at $\times 20$, and diffuse at $\times 40$, appearing more or less circular with a slight central condensation. M95, 40' to the west of M96, is half a magnitude fainter and more diffuse, with no obvious condensation, appearing elongated in the NNE–SSW direction. M96 has an apparent diameter of 2', while M95 spans about 3'. These are both SBb barred spirals. M105, 50' NNE of M96, is a class E1 giant elliptical galaxy, appearing as a faint, even circular glow 2' wide.

All three members of this group – part of the same extended, 30 million light years distant cluster of galaxies to which M65, M66 and NGC 3628 belong – fit comfortably in the field of view at $\times 40$. M95 and M96 were discovered by Méchain in 1781.

## NGC 2903
Leo    RA 09h 32.2m dec +21° 30'    mag. +9.0    Map 3

One of Leo's brighter galaxies, NGC 2903 is perhaps a surprising omission from Messier's catalog, being prominent in small telescopes. NGC 2903 is easy to find, just west of the Sickle of Leo. The mag. +4.3 star $\lambda$ Leo is the best guide: just south of $\lambda$ are a couple of 7th-magnitude stars aligned roughly E–W, and NGC 2903 is 22' south of the more easterly of the two. I find this galaxy prominent in my 80 mm f/5 refractor at $\times 40$. The galaxy appears elongated NE–SW, with a slightly brighter nuclear region offset somewhat to the south. Visually, NGC 2903 has a long axis of about 8'. Larger telescopes show mottling: there are prominent H II regions. This Sb spiral galaxy is 25 million light years away and was discovered by William Herschel in 1784.

## M51 NGC 5194 Whirlpool Galaxy
Canes Venatici    RA 13h 29.9m dec +47° 12'    mag. +8.4    Maps 1, 4

Familiar from many photographs, the Whirlpool Galaxy (M51) is reasonably bright and easy to locate, 3.5° SW of $\eta$ UMa (Alkaid), which is at the end of the Big Dipper's handle. M51 is at the SW tip of a broad isosceles triangle that it forms with Alkaid and the 5th-magnitude 24 CVn. The Whirlpool takes its popular title from drawings made by Lord Rosse during observations with his 72-inch reflector at Birr in 1845, the first to show spiral structure in what we now know to be a galaxy.

M51 is visible in $10 \times 50$ binoculars under good conditions. Any small telescope will give an idea of its shape; the combined light of M51 and its companion to the north (NGC 5195, RA 13h 30.0m, dec +47° 18', mag. +9.6) gives an elongated outline, noticeably broader to the south.

M51 (NGC 5194) itself is a face-on Sc spiral galaxy. The spiral arms are loose, and M51 has a circular angular diameter of about 6'. A 150 mm telescope shows M51 and NGC 5195 as overlapping circular hazes, each with a central condensation; the bright centers of the two galaxies are 4.5' apart, and can be distinguished even in small instruments. M51's bright northeastern spiral arm is quite prominent in larger telescopes and can, under exceptional conditions, be seen in an instrument of 100 mm aperture.

NGC 5195 is described as peculiar/irregular, and has been distorted by a cosmologically recent encounter with its larger neighbor. Although photographs and CCD images give the impression that it is linked to a spiral arm of M51 extending to the north, NGC 5195 evidently lies beyond M51.

Discovered by Méchain in 1781, M51 is comparable in actual size to the Milky Way and is 35 million light years distant. It is the largest member of a small group of galaxies which includes M63 in Canes Venatici.

## M63 NGC 5055 Sunflower Galaxy
Canes Venatici   RA 13h 15.8m dec +42° 02'   mag. +8.6   Map 4

M63 is an Sb type spiral galaxy, at the eastern end of a broad isosceles triangle it forms with α and β CVn. M63 takes its popular name from the mottled, patchy appearance of its spiral arms, which contain numerous starclouds and H II regions. The central hub, about 4' wide and elongated E–W, is the most prominent feature in smaller telescopes. M63 is just visible in binoculars, about a degree north of the 5th/6th-magnitude stars 20 and 19 CVn; in the telescopic field, a mag. +8.5 star is just west of the galaxy. M63 was discovered in June 1779 by Méchain and lies 35 million light years away. It was one of the first "spiral nebulae" to be recognized as such by Rosse.

## M101 NGC 5457
Ursa Major   RA 14h 03.2m dec +54° 21'   mag. +7.9   Maps 1, 4   Finder chart p.39

Quite easily found by a star-hop eastward along the handle of the Big Dipper from Mizar (ζ UMa, p.136), M101 is an Sc spiral galaxy presented face-on. As a result, its surface brightness is fairly low, and despite its integrated catalog magnitude of +7.9 it is a testing binocular target except under the very best conditions.

Small to medium amateur telescopes show only the bright core – about the innermost 25% – of this galaxy. The spiral arms are faint, with brighter patches and knots indicating the presence of H II regions and starclouds, some of which have their own NGC numbers. The extended spiral arms give an overall size of 26', corresponding to an actual diameter of 190,000 light years – M101 is a very large spiral galaxy!

M101 was discovered by Méchain in March 1781. Its spiral nature was first surmised by Lord Rosse in the 1840s. The most prominent member of a small group of mainly faint galaxies, M101 lies at a distance of 27 million light years.

## M64 NGC 4826 Black Eye Galaxy

Coma Berenices  RA 12h 56.7m dec +21° 41'  mag. +8.5  Map 4

Visible in 10 × 50 binoculars as slightly non-stellar under good conditions, M64 is quite easy to find, a degree NE of the mag. +4.5 star 35 Com. 35 Com itself is located about a third of the way NE from α toward γ Com at the northern tip of the triangle of stars making up the main part of Coma Berenices.

In my 80 mm refractor, M64 appears brighter than its catalog magnitude of +8.5 would suggest, and somewhat elongated SE–NW. Even at ×80, this small instrument reveals none of the structure from which the galaxy takes its popular name. A 110 mm reflector used at ×31 reveals a small, ill-defined nucleus. Higher magnifications with such an instrument begin to hint at the dark material just north of the galaxy's core; a 150 mm shows this quite clearly. M64 is a quite tightly wound Sb spiral galaxy with apparent dimensions of 9' × 5'.

M64's dark material is indicative of vigorous ongoing star formation. Discovered by Bode in 1779, M64 lies at a distance of 19 million light years and has an actual diameter of 51,000 light years. It is part of a small group which includes M94 in the neighboring constellation Canes Venatici.

## M94 NGC 4736

Canes Venatici  RA 12h 50.9m dec +41° 07'  mag. +8.2  Map 4

Lying 3° NNW of α CVn (Cor Caroli), at the apex of a flat isosceles triangle it forms with α and β CVn, M94 is quite an easy binocular object. In small telescopes it appears compact, circular and only slightly "fuzzy," with good contrast against the sky. An Sb spiral galaxy, it was discovered by Méchain in May 1781 and is at a distance of 21 million light years.

### The Virgo–Coma Cluster

The area from the NE corner of the Virgo Bowl north toward the triangle of stars that represents the tresses of Coma Berenices abounds with reasonably bright galaxies (and, indeed, a great many more NGC-listed objects which are in reach for very large – 200 mm aperture and greater – amateur telescopes). Navigating around these galaxies can be a daunting prospect, but with a little time and patience it can be very rewarding. To take in all the Messier galaxies in this region will require a minimum of a couple of hours' observing time, at least on the first attempt. In many ways it is better to work your way slowly around this region over the course of a couple of nights. I find the easiest approach is to seek out these galaxies in three separate hops, rather than try to do the whole collection at once. While the Virgo Bowl region has relatively few bright stars to use as landmarks, there are enough in the mag. +6 to +7 range for successful galaxy-hunting; low-power binocular reconnaissance is certainly a good idea before taking the plunge telescopically.

### A dip into the Virgo Bowl

| M59 NGC 4621 | | | | |
|---|---|---|---|---|
| Virgo | RA 12h 42.0m dec +11° 39' | mag. +9.6 | Map 4 | |
| **M60 NGC 4649** | | | | |
| Virgo | RA 12h 43.7m dec +11° 33' | mag. +8.8 | Map 4 | |
| **M58 NGC 4579** | | | | |
| Virgo | RA 12h 37.7m dec +11° 49' | mag. +9.7 | Map 4 | |
| **M89 NGC 4552** | | | | |
| Virgo | RA 12h 35.7m dec +12° 33' | mag. +9.8 | Map 4 | |
| **M90 NGC 4569** | | | | |
| Virgo | RA 12h 36.8m dec +13° 10' | mag. +9.5 | Map 4 | |
| Finder chart for all objects on p.63 | | | | |

As a first excursion into the Virgo Bowl, I suggest starting with the five relatively bright galaxies at its eastern end. These are easy to find,

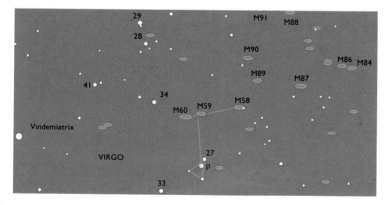

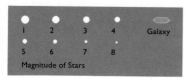

▲ Finder chart for galaxies in the eastern part of the Virgo Bowl. The mag. +4.5 star ρ Vir is a good starting point for location. This chart has an angular width of 10°. Stars are shown to limiting magnitude +8.5.

using the mag. +4.5 star ρ Vir as a guide. ρ is 4° west of the third-magnitude ε Vir (Vindemiatrix), and is at the center of a small triangle of 6th- to 7th-magnitude stars. A line from the southwestern star of this triangle through ρ extended northward by about four times the distance between them (a hop of 80′ from ρ) takes the observer straight to M59. M60 is just 25′ to the east of M59, and the two galaxies are comfortably visible together in my 80 mm refractor at ×40. M60 appears bright, with a strong central condensation, while M59 is more diffuse but with a more concentrated nucleus than M60. Particularly in averted vision, M59 seems slightly the larger. Both galaxies are giant ellipticals, with apparent diameters around 2′. M59 is class E5, while M60 is E2.

Coming west from M59 by a degree (slightly less than a field-width at ×40) takes the view to M58, a type SBb barred spiral which has a mag. +8 foreground star about 10′ to its west. In a small telescope at ×40 this galaxy is quite faint, and is best seen in averted vision. M58 appears fairly compact (about 2′ × 1′) and elongated E–W. Closer examination shows the central bar as a condensation immersed in the haze of the outer arms.

From M58, hopping another ×40 field-width, this time to the NW, takes the observer to M89, a small, 1′-diameter elliptical galaxy appearing only as a fuzzy spot. Slightly more prominent 30′ to the NE of M89 (and in the same low-power field) is M90, a spiral galaxy with a bright core region, 5′ × 2′ in size with the long axis oriented NNE–SSW.

### Coma Galaxies

| M98 NGC 4192 | | | | |
|---|---|---|---|---|
| Coma Berenices | RA 12h 13.8m dec +14° 54′ | mag. +10.1 | Map 4 | |

| M99 NGC 4254 | | | | |
|---|---|---|---|---|
| Coma Berenices | RA 12h 18.8m dec +14° 25′ | mag. +9.9 | Map 4 | |

| M100 NGC 4321 | | | | |
|---|---|---|---|---|
| Coma Berenices | RA 12h 22.6m dec +15° 47′ | mag. +9.3 | Map 4 | |

| M85 NGC 4382 | | | | |
|---|---|---|---|---|
| Coma Berenices | RA 12h 25.4m dec +18° 11′ | mag. +9.1 | Map 4 | |

| M88 NGC 4501 | | | | |
|---|---|---|---|---|
| Coma Berenices | RA 12h 32.0m dec +14° 25′ | mag. +9.6 | Map 4 | |

| M91 NGC 4548 | | | | |
|---|---|---|---|---|
| Coma Berenices | RA 12h 35.4m dec +14° 30′ | mag. +10.2 | Map 4 | |

Finder chart for all objects on p.65

The second tranche of Virgo–Coma Cluster galaxies can be picked off by coming in from Denebola, which marks Leo's tail, to the 5th-magnitude star 6 Com, 6° to its east.

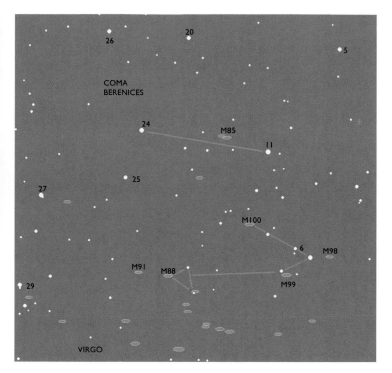

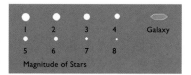

▲ *Finder chart for galaxies in Coma Berenices, to the east of Leo's "tail." The chart has an angular width of 10° and shows stars to a limiting magnitude +8.5.*

Among the faintest galaxies in this selection is M98, a fairly edge-on spiral 30′ due west of 6 Com. I barely glimpse this in averted vision in an 80 mm telescope.

Rather more prominent is M99, which lies about 40′ SE of 6 Com between a pair of 7th/8th-magnitude stars. M99 is a class Sc spiral with an apparent diameter of about 3′, showing slight central condensation.

Returning to 6 Com, a star-hop eastward through two 6th-magnitude stars takes the field to M100, a face-on barred spiral galaxy with a strong nucleus and quite bright surrounding arms, seen as an oval haze of 4′ × 3′ elongated ESE–WNW.

North from M100, between the 5th-magnitude stars 11 and 24 Com, M85 is a bright spiral galaxy, well seen in my 80 mm at ×20 as a hazy spot. A mag. +10 star foreground star appears superimposed (deep sky observers sometimes describe such objects as being "involved") to the south, while a 9th-magnitude star is nearby to the NE. At ×40, M85 appears to have a near-stellar nucleus and an overall diameter of 3'.

Coming back southward in the direction of the Virgo Bowl, return to 6 Com, then M99, and eastward 2.5° to a couple of 7th-magnitude stars lying NNE–SSW. Making an isosceles triangle with these (the southern star being its sharp point) is M88. In my 80 mm at ×20, this galaxy is quite obvious, and better still at ×40, framed between two 10th-magnitude stars (and closer to the more southerly of these). M88 appears elongated N–S and perhaps slightly brighter at its southern end. This is a type Sb spiral, with an apparent size of 5' × 2'.

Due east from M88 by 55', M91 is rather like M99 – a faint, diffuse object for small telescopes, barely glimpsed using averted vision as a circular haze at ×40 in my 80 mm.

## Three Giant Ellipticals

| M84 NGC 4374 |
|---|
| Virgo   RA 12h 25.1m dec +12° 53'   mag. +9.1   Map 4 |
| **M86 NGC 4406** |
| Virgo   RA 12h 26.2m dec +12° 57'   mag. +8.9   Map 4 |
| **M87 NGC 4486** |
| Virgo   RA 12h 30.8m dec +12° 24'   mag. +8.6   Map 4 |
| Finder chart for all objects on p.67 |

Back across the border in Virgo are three relatively easy galaxy targets, found by using ρ Vir as a starting point again. A couple of degrees west of ρ is 6th-magnitude 20 Vir, from which a couple of lines of 7th-magnitude stars extend to the NE. From the second star north in the more westerly line (about 1.5° NNE of 20 Vir), come west by 2° to a tight triangle – about 15' to a side – of 7th-magnitude stars: these are easily visible in a finder telescope. Extending the line between the two westernmost stars of the triangle NW by about 45' takes the observer to M84.

M86 lies 20' to M84's east, and both are comfortably held in a ×40 field. These objects are bright and circular; M86 is the larger, about 2' in size compared with 1.5' for M84. Both have marked central condensations, that of M84 appearing proportionally larger in relation to the galaxy's overall diameter.

ESE from this pair, about 30' from the NE corner of the triangle of stars used to locate M84 and M86, is M87, the largest and brightest of the Virgo elliptical galaxies. Like M86 and M84, M87 has a strong central condensation. Comparatively bright and easy, it has a mag. +7 star just to its north, and an apparent diameter of 2'.

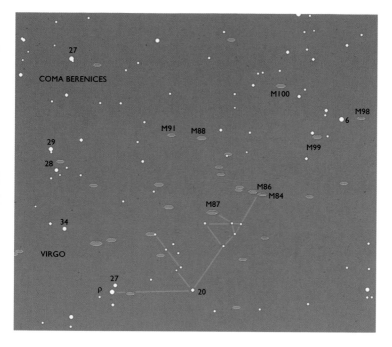

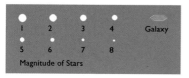

Magnitude of Stars

▲ *Finder chart for Virgo's giant elliptical galaxies. Again, a good starting point is ρ Vir. The view is 10° in angular width, and stars are shown to a limiting magnitude of +8.5.*

All three of these galaxies – a combined total of thousands of billions of stars – can be seen in the same field at ×20. M84 is a class E1 galaxy, M84 is an E3 and M87 an E0. M87 is noted for its active nucleus, which emits a jet of material (visible only in very large telescopes), and it is a strong radio source.

## M104 NGC 4594 Sombrero Galaxy
Virgo   RA 12h 40.0m dec −11° 37′   mag. +8.0   Map 4

Most amateur astronomers know of Virgo as a happy hunting ground for galaxies, many of which are concentrated at the NE side of the Bowl of the constellation in a cluster which spills over the border into the neighboring Coma Berenices. Several of the Virgo–Coma Cluster's brighter members were included in Messier's catalog and are accessible in small amateur instruments. Many of the outliers from the main concentration are well worth seeking out, too. Among the most celebrated is the Sombrero Galaxy (M104, NGC 4594), well south of the Virgo Bowl and right on the border with Corvus. The Sombrero is easy to find, using the quadrilateral of Corvus' main stars as a guide. A line through ε at the quadrilateral's SW corner extended through δ at the NE by a little more than the distance between these stars takes the observer to M104. Alternatively, come 8° due west from Spica.

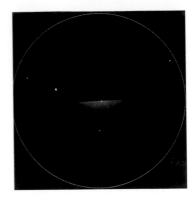

▲ A sketch of M104 (the Sombrero Galaxy) by John Lewis. It has been rendered in chalk on a black background to give an impression resembling the eyepiece view.

M104 is an Sa spiral galaxy, with a prominent nuclear bulge. It is presented only 6° from edge-on to us, and, in large telescopes particularly, material in the galaxy's spiral arms appears as an obscuring dark lane silhouetted against the nucleus. In small telescopes the nucleus is the most prominent feature.

## M49 NGC 4472
### Virgo   RA 12h 29.8m dec +08° 00′   mag. +8.4   Map 4

The first of the Virgo–Coma Cluster galaxies to be discovered, by Messier in February 1771, M49 is a giant elliptical, fairly oval in shape (class E4), with a visual diameter in medium-sized amateur telescopes of about 4′. The long axis is oriented NNW–SSE. I find M49 an easy object in 10 × 50 binoculars, appearing as a diffuse spot in averted vision. The galaxy is located at the western end of a flat, broad isosceles triangle it forms with ε and ρ Vir; in a low-power telescopic view it is seen to be midway between a pair of 6th-magnitude stars separated by about a degree. In my 80 mm refractor at ×20, M49's central condensation is almost star-like, but increasing the magnification to ×40 or ×80 reveals it to be quite extended. Some observers describe M49 as resembling a faint globular cluster. Photographically, M49 has dimensions of 8.1′ × 7.1′, corresponding to an actual size of 153,000 light years for the long axis.

## M61 NGC 4303
### Virgo   RA 12h 21.9m dec +04° 28′   mag. +9.7   Map 4

Located 3.5° SW of M49, a degree to the NNW of 5th-magnitude 16 Vir, M61 is beyond the reach of 10 × 50 binoculars, but is visible in small telescopes. In an 80 mm at ×20 it appears circular and similar in size to M49, but a good deal fainter. In the slightly darker field at ×40 it is perhaps a little more obvious, appearing fairly evenly bright across its diameter.

Larger telescopes show a central bar oriented N–S surrounded by bright spiral arms. M61 is an open spiral with a strong bar and fairly loose arms – and is presented face-on. Visually, it has an apparent diameter of 3′, while long-exposure images extend this to 6′, corresponding to an actual dimension of 110,000 light years. M61 was found in 1779 by the Italian observer Barnaba Oriani (1752–1832), just a few days ahead of Messier's independent discovery.

## M74 NGC 628
Pisces   RA 01h 36.7m dec +15° 47′   mag. +9.4   Map 7

One of the more testing galaxies in Messier's catalog, M74 is located 30′ NE of the mag. +3.6 star η Psc. This galaxy is a face-on Sc spiral, with a small, concentrated nucleus and spiral arms of very low surface brightness; it can be hard to see even in large telescopes. In a 150 mm aperture instrument I found that M74 was best seen at a relatively low

power, ×40; higher magnifications made it appear rather dim. Deep images show a circular diameter of around 11′, and reveal H II regions and starclouds as knots in the loosely wound spiral arms. Visually, I had the impression that the galaxy was elongated N–S. M74 was discovered by Méchain in September 1780, and lies 35 million light years away.

In January 2002, a supernova reaching peak mag. +12.3 was detected in M74's spiral arms. SN 2002ap was later found to be a "hypernova" – a Type I (p.160) event in which a star over 40 times the mass of the Sun exploded.

▲ The face-on spiral galaxy M74 in Pisces in a CCD image by Martin Mobberley. Supernova 2002ap is indicated by fine white lines at lower left of center.

## M77 NGC 1068
Cetus   RA 02h 42.7m dec −00° 01′   mag. +8.9   Map 7

Another rather difficult face-on spiral, the type Sb M77 is 42′ ESE of the mag. +4.1 star δ Cet, at the west of the triangle making up the whale's head. Although it has overall photographic dimensions of 7' × 6', only the innermost 2' of M77 is really visible in amateur telescopes. This relatively bright core region is surrounded by knotted, rather faint spiral arms. M77 is a source of strong radio emission, some of which seems to be associated with vigorous star formation. This is a large galaxy (170,000 light years in diameter) 60 million light years away. It was discovered in October 1780 by Méchain.

## NGC 3115 Spindle Galaxy
Sextans  RA 10h 05.2m dec −07° 43'  mag. +8.9  Map 3

The brightest example of a lenticular (S0) galaxy is NGC 3115 in Sextans, which can be found two-fifths of the way (about 5°) along a line running north from λ Hya to α Sextantis; these are both 4th-magnitude stars in the rather empty sky due south of Regulus. Under good conditions the bright nucleus of this almost edge-on object is visible in binoculars. Popularly known as the Spindle Galaxy, NGC 3115 has overall dimensions of 8' × 3'. It was discovered by William Herschel in February 1787 and lies 25 million light years away.

## M83 NGC 5236
Hydra  RA 13h 37.0m dec −29° 52' mag. +7.6  Map 4

M83 is a face-on Sc spiral galaxy showing quite low contrast, discovered by the French astronomer Nicolas de Lacaille (1713–62) during his two-year survey of the southern sky from southern Africa, begun in 1751. Although its integrated magnitude is relatively bright at +7.6, this is a testing object for observers at northerly latitudes: from southern England, it culminates barely 10° above the horizon on a spring night, and any haze will render the search fruitless. M83 lies near Hydra's border with Centaurus, 7° SSE from 3rd-magnitude γ Hya, which is 8° due east of β Crv – a lack of nearby bright field stars makes this object quite tricky to locate.

M83 has dimensions of 15' × 13', but in medium-sized amateur instruments it appears about half this size. The central regions are fairly bright, and look like an oval "bar" (though this is not classed as a barred spiral galaxy) extending from a bright core. Larger telescopes show mottled H II regions. Several supernovae have been recorded in the galaxy's openly presented spiral arms. M83 is 22 million light years away, and is part of a small galaxy group which includes NGC 5128 in Centaurus.

## NGC 253 Silver Coin
Sculptor  RA 00h 47.6m dec −25° 17'  mag. +7.8  Map 7

Discovered by Caroline Herschel during her comet-hunting on September 23, 1783, NGC 253 is a fine Sc spiral galaxy. Its nearly edge-on presentation means that although the spiral arms are loosely wound, their combined light is concentrated into a smaller area of sky, giving it a relatively high surface brightness and good contrast: under good conditions it can be seen well with binoculars, although its low elevation in the sky from Northwest Europe limits visibility. NGC 253 has overall dimensions of $25' \times 7'$, markedly elongated NE–SW. Deep images show a lot of mottling in the spiral arms, indicative of ongoing star formation. Its photographic appearance has gained it the nickname of the Silver Coin.

The galaxy is located on a rather empty part of the sky, close to the South Galactic Pole (a projection onto the celestial sphere of the Milky Way's axis). It can be found by taking a line 7.5° south from the 2nd-magnitude star β Cet (Diphda). Observers at southerly latitudes can use 4th-magnitude α Scl as an additional guide: NGC 253 lies about a third of the way north along a line from α Scl to Diphda. It is the largest and brightest member of a small galaxy group at a distance of 10 million light years, which vies with the M81/M82 cluster for the status of closest to the Local Group. Most of the 20 or so Sculptor Group members are dwarf galaxies.

## NGC 891

Andromeda   RA 02h 22.6m dec +42° 21'   mag. +9.9   Map 7

Edge-on galaxies can be difficult targets, especially in smaller instruments – as observers find with, for example, NGC 3628 in the "Leo Trio." Dark dust lanes in galaxies seen edge-on, or nearly so, leave visible only a small luminous "surface" – in a narrow sliver or pair of narrow slivers – which may be hard to pick out unless conditions are particularly dark and transparent. This is certainly the case for NGC 891 as far as users of small telescopes are concerned: a 100 mm aperture is probably the minimum for detection of this object, 5° east of the fine double star γ And (p.140).

NGC 891 is an Sb spiral galaxy, at a distance of 10 million light years. Its overall angular dimensions are $14' \times 3'$, with the long axis aligned NNE–SSW. NGC 891 was discovered in August 1783 by Caroline Herschel, a month before she found NGC 253.

## NGC 4565

Coma Berenices   RA 12h 36.3m dec +25° 59'   mag. +9.6   Map 4

Discovered by William Herschel in 1785, NGC 4565 is widely regarded as the finest edge-on galaxy for amateur observers. Located 2° due east of 17 Com – the third star SE of γ at the apex of Coma's triangle – the galaxy is fairly prominent in small instruments. In my 80 mm at ×40, it is clearly seen in averted vision, markedly elongated SE–NW

◄ *NGC 4565 is one of the finest edge-on galaxies for amateur telescopes. Visual observers can normally detect only the nuclear bulge, split into two unequal "slivers." This deep image by Gordon Rogers shows the extended disk to either side of the galaxy's nucleus.*

and about 3′ long. Small telescopes like this show only the nuclear bulge: the dark lane so well seen in photographs and CCD images demands at least a 100 mm aperture telescope. The dark lane cuts the elongated nuclear bulge unequally, with the southern part appearing larger. NGC 4565 has overall angular dimensions of 14′ × 2′ and is classed as an Sb spiral galaxy. Its true diameter is about 125,000 light years, and it lies at a distance of 31 million light years.

# 4 · GLOBULAR CLUSTERS

Of the objects that were listed by the French comet-hunter Charles Messier in his final catalog of deep sky objects published in 1781, those that bear the closest visual resemblance to comets are the globular clusters. It is no coincidence that 13 of the 45 in his earliest list (published in 1771), and indeed 28 out of today's accepted total of 109 Messier objects, turned out to be members of this class. Modern observers had a good opportunity to make a direct comparison when sixth-magnitude Comet Levy glided past the similarly bright M15 in Pegasus over a couple of nights around August 18, 1990: the comet, appearing in binoculars and small telescopes as a diffuse, tail-less ball, was described by many observers as a larger version of the globular.

As the name suggests, globular clusters are fairly compact, usually near-spherical collections of stars. Some are extremely densely packed, containing as many as a million stars in a volume less than 200 light years across. Most are found in the Galaxy's extended, outer halo, taking hundreds of millions of years to orbit the Milky Way's center of mass; some have very elliptical orbits.

The stars in globular clusters are ancient, belonging to Baade's Population II (p.47). Astronomers can measure the ages of globular clusters (and open star clusters) by examining the spectra of the individual stars, and plotting the results on a graph of color or temperature versus luminosity (the *Hertzsprung–Russell diagram*). Most stars lie in a diagonal band crossing the diagram from top left to bottom right, but the most massive stars, which consume their hydrogen fuel by fusion most rapidly, leave this so-called main sequence to join a giant branch, a horizontal band across the top of the diagram. By determining where the branch turns off from the main sequence for a given cluster, astronomers can estimate the age of its stars. For globular clusters, this turns out to be of the order of 12 billion years, making them very ancient indeed.

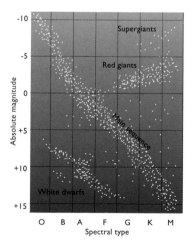

▲ In the Hertzsprung–Russell diagram, evolved red giant stars form a branch to the right of the main sequence. For a given globular cluster, the absolute magnitude at which this branch appears can be used to determine the object's age.

It is widely thought that globular clusters formed early in the history of the Universe, and that they may even represent building blocks which, through mergers, grew into the larger structures of galaxies. The globular clusters surrounding our Galaxy may therefore be leftovers from the Milky Way's formation. Some of the streams of stars that share common motions in the outer arms of the Galaxy were probably originally part of now-disrupted globular clusters.

Astronomers describe all elements heavier than hydrogen and helium as *heavy elements* or *metals*. Metals are produced by nuclear fusion deep within stars and, in the case of elements heavier than iron, in the violent explosion of very massive stars as supernovae. Thrown out into the interstellar medium, these heavy elements can become incorporated into later generations of stars. Population II stars, like those in globular clusters, have very low metallicities (metal contents), dating as they do from a time when heavy elements were not present in any significant quantities in the interstellar medium: they hadn't yet been synthesized. The low metallicity of globular clusters unfortunately destroys the science fiction cliché of spectacular star-filled skies for planets located there: the lack of heavy elements makes the occurrence of planets like ours within the confines of a globular cluster most unlikely.

◄ Associated with the Andromeda Galaxy (M31) is the globular cluster G1. It is the most massive in the Local Group, and can be seen in very large amateur telescopes.

From our terrestrial perspective in one of the Galaxy's metal-rich outer spiral arms, we see globular clusters largely congregated in the direction of the Milky Way's hub. The constellations Ophiuchus, Scorpius and Sagittarius are rich hunting grounds for these objects, and nights between May and September can be regarded as the prime globular cluster season. Conversely, in December skies, the brightest example is the relatively dim M79 in Lepus, something of an outlier from the main aggregation.

In 1917, the American astronomer Harlow Shapley (1885–1972) used the distribution of globular clusters to determine the approximate position of the Galactic center. Realizing that the globulars were in orbit around the Milky Way's heart, Shapley was able to place this, correctly, in Sagittarius.

While most globular clusters are found in the line of sight to the Galactic hub, several outliers are found in other parts of the sky. A notable example is the faint, so-called Intergalactic Tramp (NGC 2419) in Lynx, which lies an estimated 300,000 light years from the center of our Galaxy.

Globular clusters are seen around other galaxies too. A popular target for large amateur telescopes is G1, a globular cluster gravitationally bound to the Andromeda Galaxy (M31). At magnitude +13.7, G1 requires a 200 mm aperture telescope and excellent observing conditions if it is to be detected; it lies 2.5° SW of the parent galaxy's nucleus at RA 00h 32.8m, dec +39° 55'. Our Galaxy has a population of about 150 globular clusters; M31 has at least 350. Deep, professional images of the giant elliptical galaxy M87 (p.66), at the core of the Virgo–Coma Cluster of galaxies, show it to be surrounded by thousands of globulars.

Many of the Milky Way's globular clusters lie within binocular range, and many more can be seen in small amateur telescopes. The two brightest examples are easily visible to the naked eye, and are identified by their original, mistakenly stellar designations as Omega Centauri and 47 Tucanae.

Globular clusters come in a range of apparent size and brightness. A critical factor in determining ease of detection is a globular's degree of condensation (just as it is for comets). Seasoned observers use a 12-point scale to describe this feature, ranging from I (very dense and compact, almost stellar in appearance in some cases) to XII (extremely diffuse). A densely packed, star-rich cluster with a high degree of condensation will show greater contrast with the sky background and will therefore be easier to see. Rich globular clusters are more forgiving of poor sky conditions than any other type of deep sky object, and several are quite well seen from suburban locations, or on twilit or moonlit nights.

▲ *The most spectacular of the globular clusters in the northern half of the sky, M13 is a popular target for North American and European observers. This image by Robert Gendler resolves M13's outer parts into individual stars.*

Some globulars are quite easily resolved into individual stars when seen in a reasonable-sized telescope (100 mm aperture or greater); the bright Scorpius globular M4 is a good example. Others, such as the northern-hemisphere showpiece M13 in Hercules, require a little more aperture and magnification (150 mm, ×200) before they will appear as more than just hazy balls in the eyepiece. Some of the more distant objects, on the far side of the Galactic hub, are dimmed by interstellar dust in the line of sight and difficult to see at all.

Unlike loose open clusters, where the positions of individual stars are comparatively easy to draw, globulars are usually so rich that only a general impression of the stellar density can be given in a sketch. It is sensible for future reference to note the positions of field stars in any drawing.

## Selected Globular Clusters

### Omega Centauri NGC 5139
Centaurus   RA 13h 26.8m dec −47° 29′   mag. +3.5   Maps 4, 8

The brightest of all globular clusters is popularly known by the star designation originally assigned to it. Although it must surely have been known since antiquity, "discovery" of this object as nebulous is widely ascribed to Edmond Halley (1656–1742) in 1677. Clearly non-stellar to the naked eye, and bright at magnitude +3.5, ω Cen is a fine object in any instrument. European observers are just too far north to see ω Cen, but from the southern United States it is reasonably well placed. It is a favorite with observers at the northern-wintertime Florida Star Party, getting up to about 20° in the sky at 25°N latitude. The best views, naturally, are from southern-hemisphere locations, where even binoculars will hint at resolution of this magnificent globular cluster's outer regions.

In a small telescope, many more stars are resolved, giving the impression of an even sprinkling all the way down to the center. ω Cen is a fairly loose globular, with a degree of condensation of VII. Observers using large telescopes and high magnifications describe this object as breathtaking. Its large apparent size of just over 36′ – larger than the Moon's – translates to an actual diameter of 170 light years at its distance of 15,600 light years, and the total luminosity of the globular is 1.1 million times that of the Sun. Over a million stars – totaling 5 million solar masses – are crammed into this space. ω Cen is probably the largest and most luminous of the Milky Way's globular clusters, and among those associated with Local Group galaxies only G1 – bound to the Andromeda Galaxy – is thought to be larger and intrinsically brighter.

## 47 Tucanae NGC 104
Tucana   RA 00h 24.1m dec −72° 05′   mag. +4.0   Map 8

Still further south than ω Cen, and inaccessible to observers north of 18°N latitude, is the second-brightest globular cluster, 47 Tuc (NGC 104). Like ω Cen, this fuzzy "star" received a stellar designation in the seventeenth century, though its non-stellar nature was established soon after. Lacaille listed it in his catalog of southern nebulae. At mag. +4.0, 47 Tuc is only a little fainter than ω Cen, but its core region is more compact: its degree of condensation is III. Its overall diameter is an impressive 31′, translating to an actual 120 light years at its distance of 13,400 light years.

## M22 NGC 6656
Sagittarius   RA 18h 36.4m dec −23° 54′   mag. 5.1   Map 5   Finder chart p.88

The finest globular cluster visible to observers at the latitudes of North America and the British Isles is surely M22, a bright, rich, class VII object of magnitude +5.1; it is visible to the naked eye for observers in the southern hemisphere. Its southerly declination means that M22, even when best placed, is low for observers in the UK. Even in the far south of the country, this splendid object culminates at an altitude of only about 15° – it would be far better known to northern observers if

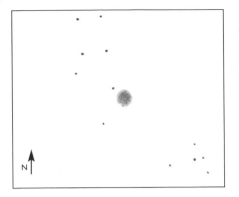

◄ *M22 in Sagittarius, as sketched by the author from the view in his 80 mm f/5 refractor at ×40. Even in this small telescope, M22 presents a mottled, partly resolved appearance.*

higher in their sky. M22 lies a couple of degrees to the NE of the 3rd-magnitude star λ Sgr, sometimes portrayed as the "lid" of the Teapot asterism made up from Sagittarius' main bright stars.

Binoculars show M22 as a large, bright hazy spot. In an 80 mm refractor, even at ×20, this globular appears mottled. When the magnification is raised to ×100, M22 fills about a third of the field – giving a rough apparent diameter of 20', significantly larger than that of the Great Globular (M13) in Hercules – and shows substantial resolution into single stars. One or two brighter stars stand out from the overall rich haze, most prominently in the cluster's SE quadrant. M22's light seems fairly evenly distributed overall, with no noticeable central condensation: pinpricks of individual stars are visible all the way down to the core.

M22 has been known since the earliest application of telescopes to astronomy in the seventeenth century. At a distance of 10,000 light years it is one of the closest globulars. It has an estimated actual diameter of 70 light years and contains around 70,000 stars.

## M28 NGC 6626

Sagittarius   RA 18h 24.5m dec −24° 52'   mag. +6.8   Map 5   Finder chart p.88

Rather overwhelmed by its bright extended neighbor M22, M28 is a compact class IV globular cluster, located a degree to the NW of λ Sgr. At mag. +6.8, it is a reasonably easy binocular object – though, like M22, it is perhaps less well known to northern-hemisphere observers than it might be thanks to its relatively low maximum elevation in the sky. M28 has a condensed core with a loose outer halo extending to an overall diameter of about 5'. It was discovered by Messier in July 1764, and lies at a distance of 19,000 light years.

## NGC 6752

Pavo   RA 19h 10.9m dec −59° 59'   mag. +5.4   Map 8

Another southern-hemisphere globular cluster, invisible above 30°N latitude, NGC 6572 in Pavo is a well-resolved class VI object at a distance of 14,000 light years. Large and bright, with an overall diameter of 20′, NGC 6752 is visible to the naked eye under good conditions, and its outer regions are readily resolved in medium-aperture amateur telescopes.

## M13 NGC 6205

Hercules   RA 16h 41.7m dec +36° 28′   mag. +5.7   Map 5

The best-known globular cluster for northern-hemisphere observers is M13, the Great Globular in Hercules. At mag. +5.7 it is a very easy object, visible without any difficulty in binoculars and finder telescopes. M13 is also easy to locate, lying on the western side of the Keystone asterism at the center of Hercules, about a third of the distance south from η Her on the line toward ζ. In the low-power field of binoculars or a finder, M13 is flanked east and west by a pair of 6th-magnitude stars. The globular itself appears as a fuzzy circular patch with a slight central condensation and an overall diameter of 17′ (more than half as wide as the Moon).

Binoculars show it as a hazy, obviously non-stellar spot. Any small telescope will enhance this impression. My 80 mm short-focus refractor shows it nicely as a mottled ball with a slightly condensed core at ×20, while the view at ×40 gives a good impression of M13's compact richness. The core appears somewhat asymmetric, with concentrations to the south and NW. At ×40, there is a hint of granularity to the globular, and on a good night in a view at ×80, even with an instrument this small, there are suggestions of resolution into individual stars in the outer regions of the cluster. The degree of condensation is V.

Few observers forget their first view of M13 in its full splendor through a reasonably large aperture telescope. At ×200 in the eyepiece of a 150 mm instrument, M13 fills the field with a swarm of bluish-green stars, all roughly the same brightness. On a night of steady seeing, individual stars can be resolved quite a way down toward the central core. The sheer mass of stars – there may be as many as a million here – presents a dazzling spectacle, and the eye and brain combine to discern patterns and dark lanes within the ball of the globular.

M13 is best viewed on evenings between April and September (late spring to early

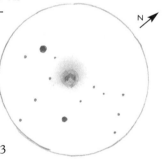

▲ The author's sketch of M13, as seen in an 80 mm f/5 refractor at ×80. M13 appears between two bright stars. The sketch can only convey some idea of the richer parts of M13, which contain far too many stars to draw individually.

autumn in North America and Europe). It was first noted by Halley in 1714, and was added to Messier's list in 1764. Its nature as a huge ball of stars was first determined by William Herschel in the 1780s. M13 lies 21,000 light years away and has an actual diameter of 140 light years, somewhat smaller than the southern hemisphere show-piece ω Centauri.

## M92 NGC 6341
### Hercules    RA 17h 17.1m dec +43° 08'    mag. +6.5    Map 5

Often overlooked as observers concentrate on its more celebrated neighbor M13, M92 is one of the finest globular clusters in the north-ern sky. Easily visible as a non-stellar object in binoculars and finder telescopes, M92 lies in a rather empty part of the sky, midway between the Keystone of Hercules and the Head of Draco.

With an overall diameter of 12', M92 is smaller than M13, and also appears more condensed in its center when examined in a small tele-scope. While its outer regions are fairly loose, M92 has a compact core, and is rated as having a degree of condensation of IV. The glob-ular is somewhat asymmetric, with the central condensation appear-ing offset to the NE when viewed in my 80 mm refractor at a magni-fication of ×80.

M92 was discovered by Bode in 1777. It lies 25,000 light years away and has an actual diameter of 80 light years, a little over half that of its famous neighbor. Its extremely low metallicity suggests that M92 is one of our Galaxy's most ancient globular clusters.

## M5 NGC 5904
### Serpens    RA 15h 18.6m dec +02° 05'    mag. +5.7    Map 4

The fifth-brightest globular cluster, M5 is easily located, lying 22' north of the 5th-magnitude star 5 Ser and 8° west of the triangle formed by α, λ and ε Ser. M5's overall magnitude of +5.7 is the equal of M13, but most observers would reckon this object to be less dense-ly packed than the Great Cluster in Hercules. Despite its class V degree of condensation, M5 can be resolved more or less down to the core in a 150 mm aperture telescope at ×200.

M5 is easily bright enough to be seen as a hazy spot in 10 × 50 binoculars and finder telescopes. Small telescopes show an object sim-ilar in size to M92, but with a larger, less concentrated center sur-rounded by a diffuse outer region.

This globular, 30,000 light years away, was originally discovered by the German astronomer Gottfried Kirch (1639–1710) in 1702, and later added to his catalog by Messier following his independent dis-covery in May 1764.

## M4 NGC 6121
Scorpius   RA 16h 23.6m dec −26° 32′   mag. +5.8   Maps 4, 5

At a distance of 6500 light years, M4 is probably the closest globular cluster to the Solar System. Easily found a degree due west of Antares, M4 is a loose (class IX) object which appears on the verge of resolution into individual stars even in 10 × 50 binoculars. A small telescope at a medium power (e.g. ×40) shows M4 as a circular, partly resolved scattering of faint stars with a diameter of about 20′. Messier found this object in May 1764, though it had earlier been reported, in 1746, by de Chésaux.

▲ M4 in Scorpius is one of the less condensed globular clusters, as shown in this photograph by Nick Hewitt.

## M80 NGC 6093
Scorpius   RA 16h 17.0m dec −22° 59′   mag. +7.3   Maps 4, 5

Near M4 in Scorpius is another, much more compact and concentrated (class II) globular cluster, M80. Seen low in southern skies for observers in North America or the British Isles, this is actually an easier object to pick out in binoculars than its neighbor. M80 has a dense core with a more-resolved periphery, extending to an overall apparent diameter of about 8′. It lies 2.5° NNW from the 3rd-magnitude σ Sco. Discovered by Messier in January 1781, it is 28,000 light years away.

## NGC 6397
Ara   RA 17h 40.7m dec −53° 40′   mag. +5.9   Maps 5, 8

M4 is rivaled in proximity to the Solar System by the southern-hemisphere globular cluster NGC 6397 in Ara, at an estimated distance of 7200 light years. A loose class IX object with a denser core, NGC 6397 has an overall diameter of 26′ in long-exposure images, but appears somewhat smaller visually. This is an easy binocular object, found 5° ENE of the wide 3rd-magnitude pairing of β and γ Ara.

## M3 NGC 5272
Canes Venatici   RA 13h 42.2m dec +28° 23′   mag. +5.9   Map 4

Outshone only by M13 among the globular clusters north of the celestial equator, M3 is large, bright and easy to see, even in binoculars. Although it lies in a fairly empty part of the northern sky, M3 is quite easy to locate, at the right angle of a triangle it forms with Arcturus and 3rd-magnitude γ Boo: alternatively, it can be found just under halfway along the line from Arcturus to Cor Caroli (α CVn). A class VI globular cluster, M3 looks grainy, even at ×40 in my 80 mm refractor. Instruments of 100 mm aperture begin to resolve stars in M3's outer regions, and large amateur telescopes show individual stars down to the core. Medium-power views give the impression of a few brighter stars standing out from the overall rich background of the cluster. A 6th-magnitude star lies about 40' away to the SW in low-power fields.

M3 was discovered by Charles Messier in 1764, and is about 27,000 light years away, making its apparent 16' diameter equivalent to a true diameter of 180 light years, into which space are crammed an estimated half a million stars.

## M53 NGC 5024
Coma Berenices   RA 13h 12.9m dec +18° 10'   mag. + 7.5   Map 4

Another fine northern-hemisphere globular, whose ready visibility rather belies its catalog magnitude of +7.5, M53 is easy to find, a degree NE of 4th-magnitude α Comae (7° NNE from the Virgo Bowl, 15° east of Arcturus). M53 shows clearly in 10 × 50 binoculars as an out-of-focus "star," surrounded by a peripheral haze. The impression of a haze around a large, bright core is reaffirmed in a small telescope at ×20; in my 80 mm at ×40, the periphery becomes grainy, while at ×80 a few of its brighter stars stand out. M53 has an overall apparent diameter of about 7' and is, like M13, a class V globular cluster in terms of its degree of condensation. A pair of nicely matched 9th-magnitude stars lies to the SW of M53 in the low-power (i.e. ×20) field. I find the outer parts of M53 still unresolved at ×100 in a 100 mm aperture telescope. M53 was discovered by Bode in 1775, and independently by Messier two years later. It is quite remote – 65,000 light years away.

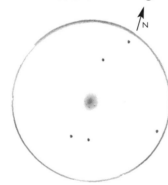

▲ The author's sketch of M53, as seen in an 80 mm f/5 refractor at ×40. The globular's more concentrated core is prominent.

A degree to the SE of M53 is a rather more difficult, and markedly different, globular cluster, NGC 5053 (RA 13h 16.4m, dec +17° 42', mag. +9.9). This is a very loose class XI object, with low luminosity. Observers using larger telescopes at low powers may be able to pull it out of the background; I have failed in several attempts to see this object, with instruments up to 100 mm aperture.

## M15 NGC 7078
Pegasus   RA 21h 30.0m dec +12° 10′   mag. +6.0   Map 6

The finest northern globular cluster on the eastern side of the Galactic hub, M15 is an easy object for August and September nights, located 4° WNW of the 2nd-magnitude star ε Peg, west of the Square of Pegasus. At mag. +6.0, M15 is an easy binocular object, appearing round and extended with a fairly even brightness across its diameter. A 100 mm aperture instrument will start to resolve the outer parts of this 7′ wide object.

M15 is interesting as an example of a globular cluster in which the phenomenon of core collapse has occurred: half the mass of this 175 light year diameter globular is found in a central condensation just over a light year across! The massive body at M15's center is a candidate for a black hole, and this globular cluster is a source of X-ray emission. M15's small dense core has been revealed by professional telescopes, including the Hubble Space Telescope. Around 20 other globular clusters associated with our Galaxy are thought to have undergone core collapse, including M30 and M62.

Discovered by the Italian-French astronomer Jean-Dominique Maraldi (1709–88) in 1746, M15 is an intrinsically luminous class IV globular cluster at a distance of 40,000 light years. Messier added it to his catalog in 1764, and William Herschel was the first to resolve it into individual stars.

## M2 NGC 7089
Aquarius   RA 21h 33.5m dec −00° 49′   mag. +6.4   Map 6

Located 4.5° due north of mag. +2.9 β Aqr, at mag. +6.4 this is one of the brighter globular clusters visible from northern latitudes. M2 is easily visible in 10 × 50 binoculars as a bright, hazy spot with a starlike core. In a small telescope, this core remains condensed, and is seen to be surrounded by a more diffuse periphery extending to an overall diameter of perhaps 10′. M2 is one of the few globulars to appear markedly elliptical, elongated SE–NW. It is a concentrated class II globular cluster containing 100,000 stars in its 175 light years diameter, and is intrinsically luminous, being readily visible despite its distance of 37,000 light years. Like M15, it was first described by Maraldi in 1746.

## M72 NGC 6981
Aquarius   RA 20h 53.5m dec −12° 32′   mag. +9.3   Map 6

Aquarius' second Messier catalog globular is rather dim in comparison with M2, at mag. +9.3. Lying 5° SE of the 4th-magnitude star ε Aqr, M72 is small (6′ in diameter) with a compact core which appears almost stellar in small telescopes. A mag. +8.5 star lies just to the east, while the asterism M73 (p.131) is a couple of degrees east of M72. Discovered by Méchain in 1780, M72 is a class IX globular cluster, fairly remote at a distance of 56,000 light years, and intrinsically quite luminous.

## M30 NGC 7099
Capricornus   RA 21h 40.4m dec −23° 11'   mag. +7.3   Map 6

Found 6° south of the 3rd-magnitude stars δ and γ Cap, at the eastern end of Capricornus, M30 is a prominent binocular object. The mag. +4.5 star 41 Cap lies 30' to the east. In 10 × 50 binoculars, M30 is an even, circular hazy patch. In a small telescope it appears markedly oval in outline, elongated E–W. The core is compact, but medium-aperture amateur instruments will resolve the outer regions of this class V object. M30 has an apparent diameter of about 4'. Like M15 in Pegasus, M30 has undergone core collapse. It has an estimated distance of 26,100 light years and was discovered in 1764 by Messier.

## M75 NGC 6864
Sagittarius   RA 20h 06.1m dec −21° 55'   mag. +8.5   Map 5

Something of an outlier from the main concentration of globular clusters around the Teapot asterism in Sagittarius, M75 lies in a rather star-poor region of sky near the border with Capricornus, 10° SW of α and β Cap. A useful guide for its location is the diamond of 5th-magnitude stars comprising ω, 59, 60 and 62 Sgr, about 15° east of the Teapot: M75 lies on a line from these stars extending toward α Cap, about 4° to their NNW.

M75 is a fairly difficult binocular target, appearing only slightly nebulous and with a stellar core in 10 × 50 binoculars. In my 80 mm refractor at ×20 the view is quite similar, but taking the magnification up to ×40 makes the core appear rather more extended and reveals some of the outer regions as a peripheral haze. At ×80, M75 appears grainy. The core seems slightly offset toward the west of center. This globular is compact, of class I, with an apparent diameter of 2.5'. Discovered by Méchain in 1780, M75 is a fairly remote 59,000 light years away.

## M55 NGC 6809
Sagittarius   RA 19h 40.0m dec −30° 58'   mag. +6.4   Map 5

Although one of the brighter Sagittarius globulars, M55 is a difficult target for observers at northerly latitudes. Its position 7° ESE of ζ Sgr at the SE of the Teapot asterism puts M55 less than 10° above the horizon at culmination from southern England, although the situation is a bit better for observers in North America: from 40°N, M55 culminates at an altitude of 20°. This is yet another globular cluster seen to best advantage from the southern hemisphere, where the Galactic center and its surroundings rise high in the sky.

M55 is bright and large, with an apparent diameter of 10′. In binoculars it has a mottled appearance, unlike the smoother light of more poorly resolved globular clusters. Even quite small telescopes, in the 80 to 100 mm aperture range, begin to resolve the outer parts of this loose, class XI object.

M55 was discovered by Lacaille during his southern Africa expedition of 1751–53, and cataloged by Messier following observations in 1778. The globular has an actual diameter of 100 light years, and lies at a distance of 17,300 light years.

| M12 NGC 6218 | | | |
|---|---|---|---|
| Ophiuchus   RA 16h 47.2m dec −01° 57′ | mag. +6.8 | Map 5 |

| M10 NGC 6254 | | | |
|---|---|---|---|
| Ophiuchus   RA 16h 57.1m dec −04° 06′ | mag. +6.6 | Map 5 |

Lying in the general direction of the Galactic hub, the large but quite faint constellation Ophiuchus is a rich hunting ground for globular clusters. Indeed, several of these objects were discovered during a particularly fruitful period in May–June 1764 by Messier: seven Ophiuchus globulars eventually ended up in his catalog. The best of these, M12 and M10, are easy binocular objects and, lying only 3° apart, are contained in the same field in a pair of 10 × 50 binoculars.

There are no obvious guide stars for finding the M12/M10 pair, but at respective magnitudes of +6.8 and +6.6, they are sufficiently obvious to be found on a first look, 10° east from the 3rd-magnitude stars δ and ε Oph; M10, the more easterly of the duo, lies a degree west of 5th-magnitude 30 Oph.

M12 is class IX, and M10 class VII, but the pair look remarkably similar. These are large, extended, more or less circular objects, each with apparent diameter of about 15′. I see them both as somewhat grainy in 10 × 50 binoculars. Even at low magnifications of ×40 and ×80, a 150 mm aperture telescope begins to resolve these globulars; at ×150 they resolve right down to the core, neither showing any marked central condensation. At high powers, a 7th-magnitude foreground star is seen to stand out against M12's southern half. M10 lies 15,000 light years away, while M12 to its NW is slightly more distant at 19,500 light years.

| M19 NGC 6273 | | | |
|---|---|---|---|
| Ophiuchus   RA 17h 02.6m dec −26° 16′ | mag. +6.7 | Map 5 |

Easily found, 8° due east of Antares, M19 is one of the more markedly non-spherical globular clusters, being somewhat elongated N–S. Its low elevation makes it a moderately testing binocular object at northern latitudes, but observers in the southern hemisphere should have no trouble picking out M19 on a reasonable night. M19 has an overall apparent diameter of about 6′, its outer fringes resolved into stars by medium-aperture amateur telescopes. My 80 mm refractor shows this globular readily, the core appearing quite concentrated at ×40. This luminous class VIII globular is 28,000 light years away, and owes its distorted appearance to its proximity to the Galactic center. Its true diameter is around 140 light years.

## M62 NGC 6266
Ophiuchus   RA 17h 01.2m dec −30° 07′   mag. +6.7   Map 5

Due south of M19 by 4° (and 8° ESE of Antares, or 4° NNE from mag. +2.3 ε Sco), M62 lies close to the Ophiuchus/Scorpius border. Its southerly declination makes this a difficult object for observers in Northwest Europe: even from the south of England, M62 culminates only about 10° above the horizon. Observers in North America have a somewhat more favorable view, but this is yet another globular cluster seen to best advantage from southerly latitudes.

M62 and M19 appear quite similar in amateur instruments. M62 is more concentrated in its center, being of class IV; like M15 in Pegasus, this globular has undergone core collapse. M62 is markedly distorted as a result of tidal forces owing to its proximity to the Galactic center, from which it is only 6100 light years distant. The core's marked offset toward east of center is clearly seen in small telescopes. M62 has an apparent visual diameter of about 6′, and its actual size is reckoned to be 100 light years; it is 20,500 light years distant. M62 was discovered by Messier in 1771.

## M9 NGC 6333
Ophiuchus   RA 17h 19.2m dec −18° 31′   mag. +7.6   Map 5

Located 4° SE from the mag. +2.4 star η Oph, M9 is reasonably prominent in binoculars and finder telescopes. This globular cluster appears somewhat oval with an elongated N–S axis, presumably a consequence of tidal stresses resulting from its proximity (5500 light years) to the Galaxy's center. Even in a 150 mm aperture telescope there is little hint of resolution into individual stars in this object: visually, only the innermost, 3′-diameter core region is detected. Long-exposure images show the cluster's full extent of 9′, corresponding to an actual diameter of about 90 light years. A 6th-magnitude star lies 45′ to the north. M9 is dimmed somewhat by interstellar dust in the line of sight toward Ophiuchus, and is 25,800 light years away. This globular was discovered by Messier in May 1784, and has a degree of condensation class VIII.

Observers using telescopes of 100 mm aperture and upward will find the rather more remote (50,000 light years) mag. +8.3 class II globular cluster NGC 6356 80′ to the NNE of M9, at RA 17h 23.6m dec −17° 49′.

## M14 NGC 6402
Ophiuchus  RA 17h 37.6m dec −03° 15′  mag. +7.6  Map 5

Quite testing for binoculars, in which it appears only slightly non-stellar, M14 is 8° due south of mag. +2.8 β Oph, or 15° east of the M12/M10 pair. Like M9, this is a fairly compact object; it resists resolution in smaller telescopes, which reveal the dense, 3′-wide core region. M14 appears smooth and circular in a small telescope. It lies inside a long triangle of mag. +6/+7 stars. Larger instruments show a slightly oval profile, with the longer axis oriented SE–NW. M14 is an intrinsically very luminous, class VIII globular cluster 33,000 light years away. It was discovered by Messier in June 1764.

## M107 NGC 6171
Ophiuchus  RA 16h 32.5m dec −13° 03′  mag. +8.1  Map 5

The faintest of the Ophiuchus globulars, M107 is found 3° SSW of 2nd-magnitude ζ Oph, and about 10° SSW from the bright M12/M10 pair. A class X object with a loose periphery and more concentrated core, M107 was discovered by Méchain in 1782. It lies just south of a tight triangle of 7th-magnitude stars. In my 80 mm refractor at ×40 this globular cluster appears grainy and elongated E–W. It has an overall photographic diameter of 10′; visually, it appears only 3′ across. M107 lies 19,000 light years away, and has an actual diameter of about 80 light years.

## NGC 4833
Musca  RA 12h 59.6m dec −70° 53′  Mag. +6.9  Map 8

A loose, class VIII globular cluster in the far southern sky, NGC 4833 was discovered by the French astronomer Lacaille during his observations from southern Africa in 1751–53. NGC 4833 lies 21,200 light years away and has an overall diameter of 13.5′.

## M79 NGC 1904
Lepus  RA 05h 24.5m dec −24° 33′  mag. +7.8  Map 2

The "Orion Quarter," in the opposite direction to the center of the Galaxy, not surprisingly contains rather few globular clusters. The brightest in this region is M79, but for observers at northerly latitudes it is a tricky object. From the south of England, for example, it culminates no more than 15° above the horizon, and at mag. +7.8 it can be

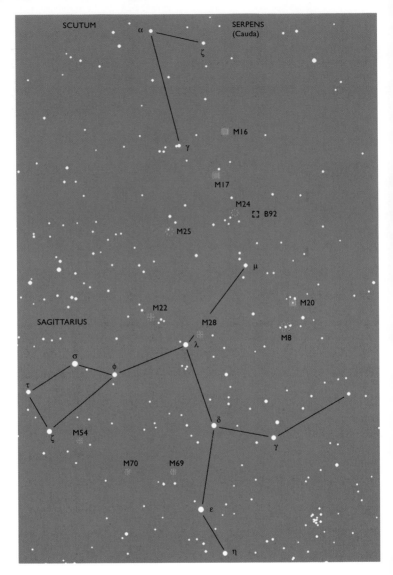

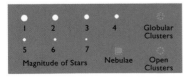

▲ Sagittarius is a rich region for globular clusters, containing seven Messier objects. Also shown in this wide-field chart (angular width 20°) are the bright nebulae M8, M20 and M17, along with M16 in Serpens (see Chapter 5).

obscured by haze on all but the best nights. Its location, at least, is quite straightforward: M79 lies 4.5° SSW of β Lep, on a line from α. A 5th-magnitude field star lies half a degree to the WSW of M79. M79 has a compressed core and a diffuse outer halo, with an overall apparent diameter of about 5′. On a good, transparent night it can be seen in 10 × 50 binoculars. Discovered by Méchain in October 1780, M79 is a class V globular cluster and is fairly remote, at 41,000 light years away. Astronomers have found that this object may have been captured recently by the Milky Way, together with the Canis Major Dwarf Galaxy.

## M68 NGC 4590
Hydra   RA 12h 39.5m dec −26° 45′   mag. +7.7   Map 4

Rather low for observers at northerly latitudes (from southern England it culminates barely 15° up), M68 is still reasonably easy to find, 3° SSE of mag. +2.7 β Crv, on a line through β from δ on the eastern side of the Corvus quadrilateral; a 5th-magnitude field star lies a degree to the SSW of M68. This is a loose, class X globular cluster with an apparent diameter of about 10′. In my 80 mm wide-field refractor at ×40, I have the impression of an asymmetrical quadrilateral outline (as opposed to the more usual circular shape for globular clusters) which is slightly broader on the western side. Discovered by Messier in 1780, M68 is 31,000 light years away.

## M56 NGC 6779
Lyra   RA 19h 16.6m dec +30° 11′   mag. +8.3   Map 5

One of the less impressive Messier globulars, M56 is found in a rich starfield midway along a line between Albireo in Cygnus and γ Lyr at the SE of the parallelogram of stars marking Lyra's body. Binoculars show M56 as a compact, circular hazy patch, while a small telescope will reveal a concentrated core surrounded by a loose periphery, and overall diameter about 5′. This class X globular cluster was discovered on January 10, 1779, by Charles Messier; on the same night he made an independent discovery of Comet C/1779 A1 Bode, which reached naked-eye visibility later in the month. As he followed the comet's passage through Virgo, Messier discovered many of the galaxies in this region which were subsequently entered in his catalog.

## M69 NGC 6637
Sagittarius   RA 18h 31.4m dec −32° 21′   mag. +7.6   Map 5   Finder chart p.88
## M70 NGC 6681
Sagittarius   RA 18h 42.2m dec −32° 18′   mag. +8.0   Map 5   Finder chart p.88

Lying as it does in the direction of the Galactic hub, Sagittarius is home to many globular clusters: seven of the 28 in Messier's catalog are found here. Among the less distinguished are the very similar pair of M69 and M70, which lie close together just south of the Teapot. Their proximity in the sky reflects the real situation: M69 and M70 are separated by only 1800 light years, and both are close to the center of the Galaxy. M69 has a distance of 6200 light years from the Galactic center, and is 29,700 light years from us; M70 is 29,300 light years away.

M69, the more westerly of the duo, is 2° NE of the mag. +1.8 star ε Sgr. At mag. +7.6, M69 is a testing object in small instruments for observers at more northerly latitudes. Its apparent diameter is about 4′, and only very large amateur telescopes reveal much more than a hazy circle of light.

M70, midway between ε and mag. +2.6 ζ Sgr, is also, at mag. +8.0, rather difficult for small telescopes, and is similar in size and appearance to its neighbor.

Both these globulars are class V objects. M70 has undergone core collapse, and is slightly smaller (68 light years diameter) than M69 (85 light years). M69 was first seen by Lacaille during his 1751–53 sojourn in southern Africa, while the discovery of M70 was made by Messier in 1780.

## M54 NGC 6715

Sagittarius   RA 18h 55.1m dec −30° 29′   mag. +7.6   Map 5   Finder chart p.88

Lying 1.5° WSW of mag. +2.6 ζ Sgr, at the SW of the Teapot, M54 is an intrinsically very luminous globular cluster (850,000 times the Sun's luminosity, outshone only by ω Cen). This object has been recently captured, along with its parent – the 80,000 light years distant Sagittarius Dwarf Elliptical Galaxy, discovered in 1994. M54 itself is 87,000 light years away.

Thanks to its large distance, the mag. +7.6 M54 has an apparent angular diameter of only about 2′, and small amateur telescopes reveal only the core of this fairly compact class II object: in binoculars and finder telescopes, M54 is almost star-like in appearance. A 100 mm aperture telescope at high magnification ($\times$150) will show some mottling, but this distant globular is beyond the limits of resolution in small instruments. M54 was discovered by Messier in July 1778.

## M71 NGC 6838

Sagitta   RA 19h 53.8m dec +18° 47′   mag. +8.3   Maps 5, 6

Midway along the length of the eastward-pointing shaft of Sagitta's arrow, formed by the stars δ and γ Sge, lies a hazy 8th-magnitude spot visible in 10 × 50 binoculars on a good night. Higher magnifications in a small telescope reveal a circular mass of stars, largely unresolved, and perhaps 6′ in diameter, standing out in a rich Milky Way starfield.

In a 114 mm reflector at ×31, M71 appears circular with a condensed center (class X–XI), and with one or two brighter stars standing out from the largely unresolved mass. An 8th-magnitude field star is quite prominent on the western edge.

The discovery of M71 is credited to Méchain, in June 1780, though it may have been seen earlier by Maraldi. For some years, astronomers debated whether this object was a globular cluster or a densely populated open cluster. The Hertzsprung–Russell diagram for stars in M71 confirms it to be a globular, and its distance of 18,000 light years is similar to that of many other globulars.

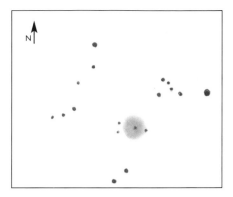

▲ *M71 as sketched by the author from the view in a 114 mm reflector at ×31. This object shows a concentrated core and a loose periphery.*

## Some Challenging Globular Clusters

### NGC 6712
Scutum   RA 18h 53.1m dec −08° 42′   mag. +8.2   Map 5

Despite its fairly bright catalog magnitude of +8.2, NGC 6712 is a tricky object for small telescopes, being compact (not much more than 3′–4′ in diameter) and rather lost in a rich starfield near the heart of the Milky Way, in Scutum, 2.5° SE from the splendid open cluster M11 (p.128). I have seen this globular in my 80 mm widefield refractor: barely distinguishable at ×20, it becomes more obvious at ×40, while at ×80 NGC 6712 appears as a circular smudge of light with some brighter stars superimposed. The globular is rather more prominent in a 125 mm aperture reflector. A triangle of 9th-magnitude stars lies just to the east of this class IX object, which is 22,500 light years away.

### NGC 6934
Delphinus   RA 20h 34.2m dec +07° 24′   mag. +8.7   Maps 5, 6
### NGC 7006
Delphinus   RA 21h 01.4m dec +16° 12′   mag. +10.5   Map 6

Delphinus is home to two challenging globular clusters, one of them remarkable for being one of the most distant globulars associated with our Galaxy.

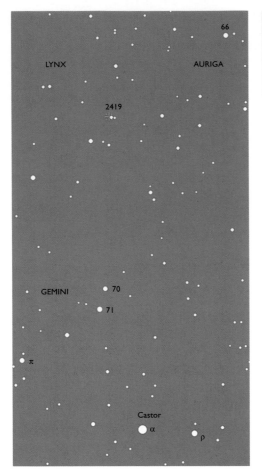

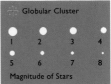

◄ Finder chart for the Intergalactic Tramp NGC 2419 in the dim constellation Lynx. North from Castor, NGC 2419 lies immediately east of a pair of 8th-magnitude stars. This chart shows field stars to limiting magnitude +8.5 and is 5.25° wide.

NGC 6934, the closer of the two, is a class VIII globular located 4° south of ε Del, which marks the Dolphin's tail. In my 80 mm refractor, this object was more obvious than I expected, appearing obviously non-stellar at ×20, and located just north of a quadrilateral of mag. +7 to +8 stars. At ×40 and ×80, NGC 6934 appears only slightly condensed at the center, with a diameter of 2′. A 9th-magnitude star is just to the west.

NGC 7006, at a distance of 150,000 light years from the Galactic center, is from our location in the Galaxy, 185,000 light years away, and because we view it through a large amount of interstellar material its light is considerably dimmed. This class I globular lies 3° east of γ Del, which marks the Dolphin's nose. NGC 7006 is framed by an isosceles triangle of stars, about a degree in its longer dimension. It is just visible in an 80 mm aperture telescope at ×40, with averted vision. To find this remote outlier in the Galactic halo, the most transparent conditions are required.

## NGC 2419 *Intergalactic Tramp*

Lynx    RA 07h 38.1m dec +38° 53′    Mag +10.3    Maps 2, 3    Finder chart p.92

Truly distant and very challenging is the Intergalactic Tramp (NGC 2419), a globular cluster which lies 300,000 light years from both the Galactic center and the Earth, in the faint northern constellation Lynx. This class II object is farther from the center of the Milky Way than are the satellite galaxies, the Magellanic Clouds. NGC 2419 is seen in the opposite half of the sky to that where most globular clusters appear to congregate. This object is a little under 7° due north of Castor, and is located just east of a pair of 8th-magnitude stars. At mag. +10.3, and with a diameter of about 2′, NGC 2419 is a very testing object for smaller amateur telescopes, requiring at least a 100 mm aperture. It was discovered by William Herschel in 1788.

# 5 · DIFFUSE NEBULAE

A nebula is, from the Latin, a "cloud" in the night sky. At one time this description would have been considered adequate for a whole variety of objects, including spiral galaxies – which were called "spiral nebulae" until their nature as remote equivalents of the Milky Way, comprised of hundreds of billions of stars, became understood in the 1930s. Nowadays the term "nebula" is restricted to a cloud of gas and dust in interstellar space. Many of the nebulae we observe are in regions where star formation is in progress, which tend to be found along the Galaxy's spiral arms close to the plane of the Milky Way. Diffuse, gaseous nebulae are distinct from the planetary nebulae (Chapter 8), which form at a late stage in the evolutionary history of Sun-like stars, and the tattered filaments of supernova remnants (Chapter 9). Several good, bright nebulae are well within the reach of modest amateur astronomers' equipment and a couple are even visible with the naked eye.

The ability of larger telescopes to resolve some apparently nebulous objects into individual stars led at one time to the supposition that

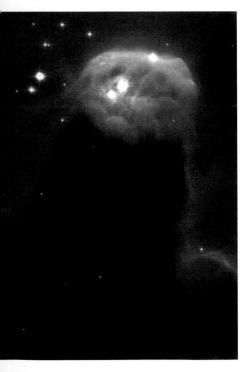

most nebulae were simply star-clouds at such great distances that the light of the constituent stars merged into a uniform haze. This notion was dispelled when observations by the English amateur astronomer William Huggins (1824–1910) in the 1860s demonstrated that several bright nebulae have *emission spectra* characteristic of gas in an excited state: unlike the *continuous spectra* of stars, nebulae emit light only at a limited number of distinct wavelengths.

True nebulae come in three principal types: bright *emission nebulae*, which are found in star-

◀ The Cone Nebula in Monoceros, imaged by the Hubble Space Telescope, is a star-forming region. The dark cone contrasts with the H II emission nebulosity which abounds in this part of the sky.

forming regions and shine as a result of excitation; *dark nebulae*, whose gas is not excited and reveals their presence by obscuring the light of background objects; and *reflection nebulae*, whose dust is illuminated by the light of nearby stars.

Emission nebulae are often described as *H II regions*. They consist of hydrogen, and are pervaded by short-wavelength, high-energy extreme ultraviolet (UV) radiation from hot young stars that have formed in the midst of the nebulosity. Extreme UV strips the single electron in a hydrogen atom from the nucleus (a proton) – in other words, the hydrogen is ionized. In an H II region, large numbers of free, positively charged protons and negatively charged electrons exist together. Occasionally, a naked proton will capture an electron, briefly becoming a neutral hydrogen atom (H I). Only certain permitted orbits – energy levels – around the proton can be occupied by a captured electron. As the electron drops toward its optimal low-energy orbit, it emits light at a distinct wavelength. H II regions thus shine strongly in red light, at a wavelength of 656.3 nm (nanometers, 1 nm = $10^{-6}$ mm). This so-called hydrogen-$\alpha$ emission is responsible for the pronounced red tint in long-exposure color photographs of objects such as the Orion Nebula or the Trifid Nebula in Sagittarius. Hydrogen-$\beta$ emission from such nebulae is also usually strong. Other emissions, such as the greenish light from doubly ionized oxygen (O III), are also found.

In addition to abundant hydrogen, nebulae contain dust particles, contributing typically about 1% of their mass. The dust consists largely of tiny grains of graphite and silicate, typically 10 nm in size, synthesized in past generations of stars. These dust grains can scatter starlight in much the same way as a fine haze of cigarette smoke scatters light in a room. Just as scattering of essentially yellowish sunlight by molecules in the Earth's atmosphere turns the sky blue, so reflection nebulae – which shine by scattering starlight from dust particles in otherwise dark interstellar clouds – often appear bluish in long-exposure images; the nebulosity associated with the Pleiades (p.114) is a familiar example.

In a typical nebula there may be 1 kg of dust grains spread through a volume of a million cubic kilometers, with an average spacing of about 10 cm between grains. Gas densities of 100–100,000 particles per cubic centimeter are found in nebulae; for comparison, the Earth's atmosphere at sea level contains $10^{19}$ particles/cm$^3$ – a factor of at least a hundred trillion denser.

Dark nebulae are just that! Good examples visible to the naked eye include the material that splits the northern Milky Way in Cygnus and Aquila into two branches (the Cygnus Rift), and the Coalsack in the southern constellation Crux. Others are prominent in the telescope as dark, obscuring features in the foreground of bright emission nebulae

◀ Dark nebulae are sometimes seen as "voids" in an otherwise star-rich field.

– the Fish's Mouth in the Orion Nebula is a good example. The Horsehead Nebula in Orion – the subject of many published long-exposure images, but very difficult to see by eye except with a very large telescope under optimal conditions – is perhaps the most familiar telescopic dark nebula.

Unlike the hot, ionized gas in emission nebulae, the hydrogen in dark nebulae is cold and in its molecular form ($H_2$). Dark nebulae have also been found to contain molecules such as carbon monoxide and more complex organic species such as formaldehyde. Characteristic microwave emissions from these molecules can be detected with radio telescopes, and used to trace the extent of nebular complexes in the Galaxy.

In addition to the giant molecular clouds, more compact dark nebulae are found in the form of *Bok globules* (named for the Dutch-American astronomer Bart J. Bok, 1906–83), which are typically a couple of light years in size and mark sites of star formation. Prominent Bok globules have been imaged in the Trifid and Rosette nebulae, but these are tricky targets for the amateur visual observer.

Dark nebulae cause reddening of light from stars beyond, rather as the Earth's atmosphere, scattering blue light more than red light from the setting Sun, creates red sunsets. They can also cause dimming, or *extinction*, of starlight, by over 20 magnitudes for stars in the direction of the Galactic center.

H II regions and dark nebulae are also seen in external galaxies, especially those relatively nearby. The face-on M33 in Triangulum (p.55), for example, has several bright H II regions which are sufficiently bright to have been given their own entries in the *New General Catalogue*. Dark nebulosity is most prominent in edge-on (or nearly so) galaxies such as NGC 4565 (p.71) and the Sombrero Galaxy (M104; p.67).

While some emission nebulae such as M42 and M8 (the Lagoon Nebula, in Sagittarius) are quite bright and prominent, most are fairly faint, extended objects with low contrast against the sky: observing them requires transparent conditions and the absence of moonlight or artificial light pollution. Large telescopic apertures and low magnifications are preferable for some extended objects.

Little, if any, color will generally be apparent in nebulae. The eye certainly won't detect the red of H II emission that is so prominent on long-exposure photographs. Where the O III emission is strong, some nebulae may appear slightly greenish. Use of O III and UHC filters can sometimes reveal interesting detail in emission nebulae; light pollution reduction filters which exclude mercury, sodium and neon emissions can enhance views of reflection nebulae. Filtered views may differ markedly from those in unfiltered light, while averted vision will certainly help in detecting some details.

Drawing nebulae can be a challenge: the eye can pick up subtle brightness variations across the width of the Orion Nebula, for example, but recording these on paper is difficult. Drawings showing the field stars and the target object's outline are worth making: some nebulae will appear more extensive on some nights than on others, depending on sky conditions.

Dark nebulae in silhouette against bright emission nebulosity can be quite obvious, and drawings of their outline are, again, worth making.

▶ *Small dark nebulae, known as Bok globules, are sites of ongoing star formation in H II regions.*

Where seen against a stellar background, however, the extent or even the presence of dark nebulae can be harder to discern. Wide-field views can help: I have, on summer nights of exceptional transparency, seen dark nebulosity as mottling in the Scutum Star Cloud with 10 × 50 binoculars, for example.

Most reflection nebulae are rather faint and, especially if associated with relatively bright stars, can be difficult to pick out. The delicate filaments around the Pleiades, for instance, are prominent on long-exposure images, but tend to get lost in the glare of starlight in small amateur instruments. Moving the telescope so that the star whose light is being reflected is out of the field of view can help.

## Selected Diffuse Nebulae

### Emission Nebulae

**NGC 3372 Eta Carinae Nebula**

Carina   RA 10h 43.8m dec −59° 52'   mag. +1.0   Map 8

A prominent 2°-diameter glow in the Milky Way west of Crux, the Eta Carinae Nebula is a large H II region (200 light years across), 8000 light years away. Named for the peculiar star embedded within it, the nebula has an integrated magnitude of +1.0, making it the brightest object of its kind. It is principally a southern-hemisphere object, below the horizon at latitudes north of 30°N. The nebula was first described by Lacaille, observing from southern Africa in 1751–53.

Several other stars lie within the nebulosity (cataloged as NGC 3372), but the most noteworthy is η Car itself. This is a very massive double star (perhaps 80 solar masses in total) which astronomers believe to be on the verge of exploding as a supernova – it will probably do so within the next 100,000 years. The star was of 4th magnitude when first cataloged (by Halley) in 1677, and has varied erratically in brightness since. In particular, a concerted period of brightening in the early nineteenth century saw it reach mag. −1 in 1843, at which time it was the second-brightest star in the night sky, outshone only by Sirius. η Car subsequently faded and is currently around 6th magnitude.

The large patch of nebulosity comprising NGC 3372 is split into two by a V-shaped dark lane. The nebula's brightest part is the 3' × 11' Keyhole Nebula, so named by John Herschel (1792–1871). η Car itself is surrounded by the 30″-diameter Homunculus Nebula, which is visible in small instruments. The Homunculus is described as markedly orange by observers, and consists of material ejected by the star during the 1800s.

| M42 NGC 1976 Orion Nebula | | | | |
|---|---|---|---|---|
| Orion | RA 05h 35.4m dec −05° 27′ | mag. +4.0 | Map 2 |
| **M43 NGC 1982** | | | | |
| Orion | RA 05h 35.6m dec −05° 16′ | mag. +5.0 | Map 2 |

Its location 5° south of the celestial equator, 4° south of ε Ori (the central star in Orion's Belt), makes the Orion Nebula (M42) a showpiece accessible to observers in either hemisphere. Rising high in the December/January midnight sky, this is the signature object for the northern winter/southern summer. At magnitude +4, M42 is visible to the naked eye under reasonable conditions. Noticeably non-stellar even without optical aid, the Orion Nebula was originally given the designation θ Ori: this is now applied to the splendid multiple star at the nebula's heart, the Trapezium (p.141).

Binoculars reveal the central half-degree or so of the main triangle of nebulosity comprising M42. M43, part of the same vast cloud of gas and dust, is visible as a small detached circular patch just to the north of M42 (it was listed separately from the main nebula by Messier in his first catalog in 1771). Even a quick, casual glance with binoculars will show these principal features, along with many fairly bright – mag. +6 to +8 – stars superimposed on the nebulosity. It is worth taking a longer look, however, with steadily mounted binoculars: even 10 × 50s will begin, on careful examination, to show some of the fainter nebulous extensions east and west of the core, bringing out the full angular extent of 85′ × 60′. M42 is a flattened wedge with the long side – to the north – aligned more or less E–W.

In any telescope the Orion Nebula is an exquisite sight. At ×20 in my wide-field 80 mm f/5 refractor, for instance, M42's brighter parts fill about half the 2.6° field. Faint tendrils of nebulosity extend away from the main wedge, and the tight Trapezium multiple star system stands out at the end of the intruding dark bay of the Fish's Mouth – a region of dark nebulosity apparently curling round in front of the bright Orion Nebula as an extension of the vast Orion Molecular Cloud which lies beyond. Often washed out in long-exposure images, there is a lot of fine detail in the nebulosity, taking the form of fairly subtle brightness variations. The northern edge of M42 forms quite

▲ The author's sketch of M42 and M43, filling the field of view in an 80 mm f/5 refractor at ×40. The dark bay of the Fish's Mouth, with the Trapezium quadruple star at its end, intrudes onto M42.

a pronounced "bar" of brighter nebulosity, while the extension to its east, particularly, curls delicately round to the south. These details look better still at ×40, the nebula filling the field of view and the Trapezium really standing out as a tight quadrilateral of stellar pinpoints against the hazy backdrop.

In addition to its hydrogen-α emission, M42 shines strongly in O III, and use of an O III filter, even on an instrument as small as 80 mm aperture, will bring out still more of the subtle "rippled" structure in the nebula, while also enhancing the contrast between the bright emission region and the Fish's Mouth.

Views in 100 mm and larger aperture instruments are simply breathtaking, showing so much fine detail that drawing the Orion Nebula becomes a daunting task indeed. In larger telescopes, the strong O III emission lends M42 a greenish cast to the eye.

About a degree to the north of M43, small telescopes will show the often overlooked reflection nebula NGC 1977 (RA 05h 35.5m, dec −04° 52′), near the mag. +4.6 star 42 Ori. Sometimes called the Running Man Nebula, NGC 1977 is elongated E–W and has a diameter of about 20′. Also in the vicinity is an attractive open cluster for small telescopes and binoculars, NGC 1981 (RA 05h 35.2m, dec −04° 26′, mag. +4.6). Covering 25′, it contains about 20 mag. +5 to +7 stars.

The Orion Nebula is only part of a vast complex of gas and dust spanning the whole constellation. We see the bright M42/M43 nebulosity because stars which have formed here recently – including those in the Trapezium – cause it to glow: this is an H II region. Eventually, radiation pressure from the stars embedded in the Orion Nebula will clear away the surrounding gas and dust, and in a few million years a bright young star cluster will take its place. Many nebular variables (p.107) are found among the stars of the Orion Nebula, and here also the Hubble Space Telescope has imaged "proplyds" – protoplanetary disks, other solar systems in the making. The bright Orion Nebula has a diameter of about 30 light years, and lies at a distance of 1600 light years in the next spiral arm out from our own in the Galaxy.

The Orion Nebula is one of the most spectacular sights in the night sky, and rewards inspection with any sort of optical aid from binoculars up to very large telescopes. Even observers who have seen it a great many times will always return for another look.

## M8 NGC 6523 Lagoon Nebula
Sagittarius    RA 18h 03.8m dec −24° 21′    mag. +6.0    Map 5

The rich Milky Way starfields in the direction of Sagittarius are graced with some of the best bright nebulae in the sky. Most prominent of these is M8, the Lagoon Nebula, 5° WNW of λ Sgr, the northernmost star of the Teapot. The Lagoon is bright and easily visible in binoculars; observers at southerly latitudes, where this object gets high in the sky, can see it with the naked eye under good conditions. In 10 × 50 binoculars, the Lagoon is a bright haze with hints of embedded faint stars, elongated E–W. Its overall span is 45′ × 30′: the longer axis is 1.5 times the Moon's apparent diameter.

Even a small telescope in the 60 to 80 mm aperture class begins to show some detail in this fine object. Low powers are sufficient to split M8 into two "lobes" of nebulosity, separated by a dark lane. The eastern half of M8 contains an open star cluster, NGC 6530, containing about a hundred members, including hot, young stars of spectral type O which illuminate the nebula. The nebulosity in the eastern lobe is thinner than on the west, giving M8 an asymmetric appearance.

▲ A sketch of the Lagoon Nebula by Alan Dowdell, using chalk on a black background to reproduce the telescopic impression.

Large telescopes show fine detail in the nebulosity, including some Bok globules. A condensed, bright portion of M8's brighter, western lobe was named the Hourglass by John Herschel, and is a site of star formation.

M8 is 5200 light years away, and has an actual size of perhaps 60 × 40 light years, making it considerably larger than the Orion Nebula (M42). The Lagoon is probably part of a single vast nebular complex which also includes the Trifid Nebula (M20) just a couple of degrees to its NW.

## M20 NGC 6514 Trifid Nebula
Sagittarius   RA 18h 02.3m dec −23° 02′   mag. +6.3   Map 5

The Trifid Nebula takes its popular name from the three dark lanes that split up the roughly circular emission nebula, 28′ in diameter, that is its most prominent feature. This is an H II region, probably part of the same complex to which the nearby Lagoon Nebula belongs. The emission nebula is surrounded by reflection nebulosity which shows bluish (in contrast to the red of hydrogen-α) on long-exposure color images, and is most conspicuous to the north.

M20 is listed in some catalogs as being as faint as magnitude +9, but its ready visibility in 10 × 50 binoculars would suggest the alternative 6th-magnitude rating to be more realistic. I have always found this an easy binocular object. In 10 × 50s it appears as an obviously nebulous extended spot with a star-like core. The Trifid and Lagoon nebulae are easily contained in the same binocular field.

Small telescopes will enlarge M20's emission nebulosity, and at ×40 the central region appears less concentrated than it does in binoculars. Hints of the reflection nebulosity to the north are apparent in an 80 mm aperture telescope, but an instrument of at least 100 mm aperture is needed to reveal the dark lanes. Very large amateur telescopes can show Bok globules in this nebulosity, where stars are still being formed.

M20 lies around 5200 light years away, on the next spiral arm inward from us in the direction of the Galactic center. M20 has a true diameter of 40 light years, slightly larger than M42 in Orion. It was discovered by Messier in June 1764.

## M17 NGC 6618 Swan Nebula
Sagittarius   RA 18h 20.8m dec −16° 11′   mag. +6.0   Map 5

Located about 7° NNE of the Trifid, the Swan Nebula (M17), together with the Eagle Nebula (M16; see below), appears to be part of another large complex of gas, dust and newly formed stars, slightly farther round the same spiral arm, at a distance of 7000 light years. Under good conditions, the Swan Nebula is a naked-eye object for observers at southern latitudes, and is easily seen in 10 × 50 binoculars.

Binoculars and small telescopes show M17 as an E–W aligned bar of nebulosity 20′ long. Even at a low power (e.g. ×20), a 60 to 80 mm refractor will show the hook of fainter nebulosity curling off north and west from the middle of the bar, forming the neck and head of the Swan. Based on this feature, M17 has alternative popular names of the Omega Nebula and Horseshoe Nebula. It is illuminated by a small cluster of stars enshrouded in it and hidden from direct view.

My most memorable view of this object thus far has surely been through a pair of 25 × 100 binoculars in July 2001, when the impression was of a wisp of cotton suspended against the rich star background: presumably as a result of the binocular view, the scene seemed convincingly three-dimensional. M17 was discovered in 1745/46 by de Chésaux, who also found M16 at around the same time.

## M16 IC 4703/NGC 6611 Eagle Nebula
Serpens   RA 18h 18.8m dec −13° 47′   mag. +6.0   Map 5

M16, the Eagle Nebula, is perhaps best known as the location of the "Pillars of Creation" imaged by the Hubble Space Telescope in 1995. While these lie beyond reach of small amateur telescopes (an aperture of 250 mm is considered the minimum necessary to view them), the nebula (IC 4703, an H II region) and the cluster (NGC 6611) are easy objects in binoculars on any reasonably dark night. M16 is easy to find, a couple of degrees NNW of M17, and covers an area of about 35′ × 28′. Much of the nebulosity is quite faint, but the view can be enhanced by using a UHC filter. See also p.14.

## NGC 2070 Tarantula Nebula
Dorado   RA 05h 38.7m dec −69° 06′   mag. +8.0(?)   Map 8

Part of the Large Magellanic Cloud at a distance of 179,000 light years, the Tarantula Nebula is a truly huge star-forming H II region over 900 light years across (more than 30 times the size of the Orion Nebula, M42!). It was near here that Supernova 1987A exploded, and deep professional images of the Tarantula's environs show numerous remnants of past supernovae – the eastern part of the LMC has spawned many short-lived massive stars.

Most catalogs give a magnitude of +8.0 for the Tarantula Nebula, which is very much in conflict with its easy naked-eye visibility for southern-hemisphere observers, and with its past stellar designation of 30 Doradus. Visual observers' reports suggest a more realistic 3rd magnitude. In 10 × 50 binoculars, the nebula is seen to extend over an area of 40′ × 25′. "Spidery" filaments of nebulosity and dark lanes are well seen in a 150 mm aperture telescope, and the Tarantula will show plenty of further detail if a UHC or an O III filter is used.

At the Tarantula's center is B136, a cluster of hot, luminous massive young stars. Strong stellar winds from these stars are responsible for shaping the nebula. Lacaille is credited with the first observation of the object, during his southern African expedition of 1751–53.

## NGC 1499 California Nebula
Perseus   RA 04h 00.7m dec +36° 37′   mag. +9   Maps 2, 7

NGC 1499 is nicknamed the California Nebula from the resemblance of its outline to the American state. It lies just north of the mag. +4.0 star ξ Per. ξ is part of a recently formed stellar association (p.111) in southern Perseus, about 1000–1600 light years away, and is a hot star of spectral type O whose ultraviolet radiation stimulates NGC 1499's emission. The nebula has sizable angular dimensions of 160′ × 40′, but is a tricky visual target because of its low surface brightness. From very dark locations it is visible in binoculars and small telescopes at low magnifications; using a UHC filter will certainly make this object easier to see. The long axis runs ESE–WNW. The California Nebula was discovered by Edward Emerson Barnard (1857–1923) in 1884/5. NGC 1499 is remarkably easy to capture, given its visual faintness, on ISO 400 to 800 film; red-sensitive emulsions pick it up in wide-field exposures of only 20 seconds at f/2

▲ *The North America Nebula in Cygnus* *usually best recorded photographically,*
*is a large, visually low-contrast H II region* *as here by Ian King.*

## NGC 7000 North America Nebula
### Cygnus   RA 20h 58.8m dec +44° 20'   mag. +5(?)   Map 6

Under good conditions, the triangular outline of the North America
Nebula (NGC 7000) can be seen with the naked eye as an enhance-
ment in the rich northern Milky Way 3° east of Deneb in Cygnus. The
nebula takes its name from the similarity of its profile – right down to
a dark bay equivalent to the Gulf of Mexico at its SW – to that of con-
tinental North America in long-exposure images.

NGC 7000 is large – 120' × 100', elongated N–S – and best seen
with low powers: 10 × 50 binoculars (under a dark sky) or a small telescope (with a UHC filter). The North America Nebula is an H II
region with strong hydrogen-α emission and lies 3000 light years away.

## NGC 2237–2239, NGC 2246 Rosette Nebula
### Monoceros   RA 06h 32.3m dec +05° 03'   mag. +10   Map 2

The Rosette Nebula in Monoceros is a prime example of how deep
sky objects can be a good deal less spectacular visually than they
appear in photographs. Located a couple of degrees east of the 4th-
magnitude ε Mon (the last in the curved chain of four stars running
SW from the foot of Pollux in Gemini), the Rosette has an overall
angular diameter of about a degree, but is of rather low surface

brightness. Although it shows up well in fairly short wide-field photographic exposures on red-sensitive film, the nebulosity in this H II region is difficult visually, except with the use of a UHC filter and/or a large-aperture telescopes.

Segments of the nebulosity have been given their own NGC listings: NGC 2237, 2238, 2239 and 2246. At the center lies the fine bright star cluster NGC 2244 (p.123). The combined stellar winds from these hot, young spectral type O and B stars have cleared a "cavity" about 20′ across at the Rosette's heart. The presence of Bok globules among the nebulosity indicates that star-formation is still going on here. The Rosette Nebula is at a distance of 2600 light years and has an actual diameter of about 130 light years.

## Reflection Nebulae

### M78 NGC 2067/2068
Orion   RA 05h 46.7m dec +00° 03′   mag. +8   Map 2   Finder chart p.106

Part of the great complex of nebulosity in Orion, 1600 light years away, M78 is found about 2° NE of ζ Ori (Alnitak), the easternmost of the three stars that represent Orion's Belt. Whereas the Orion Nebula glows with emission from hydrogen excited by stellar ultraviolet light, M78 shines by reflected starlight. M78 is, indeed, the brightest example of a reflection nebula – though at mag. +8.0 it is hardly conspicuous. In 10 × 50 binoculars it can be seen as a faint hazy patch, but it is far from easy even under good conditions.

Small telescopes show M78 reasonably well. In my 80 mm wide-field refractor, it is quite obvious at ×20, while the ×40 view shows two lobes of nebulosity, each with a faint star embedded in it. In angular size, M78 is approximately 8′ × 6′. M78's two lobes are given separate NGC entries, and are illuminated by two recently

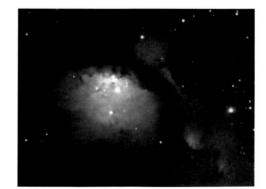

▶ M78 in Orion, here imaged by Gordon Rogers, is the brightest reflection nebula.

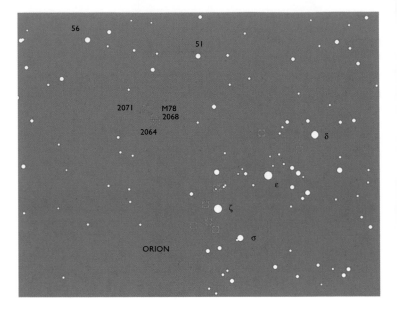

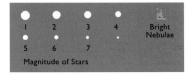

▲ *Finder chart for M78, which lies northeast of the stars that make up Orion's Belt. The chart has an angular width of 8° and shows stars to limiting magnitude +8.5.*

formed giant spectral type B stars of 10th magnitude. M78 was discovered by Méchain in 1789, and has an actual diameter of about 4 light years.

An interesting discovery in the vicinity of M78 came in January 2004, when Jay McNeil, an amateur astronomer based in Kentucky, USA, took some CCD images through a 76 mm refractor. Examination of the results showed a "new" patch of nebulosity, absent on reference images from the 1949–56 Palomar Observatory Sky Survey, at RA 05h 46.2m, dec +00° 08′ (15′ SW of M78). This faint (mag. +15 to +16) object, soon christened "McNeil's Nebula," has dimensions of about 30″ × 60″. Checks through archival photographs of M78 and its environs have turned up "prediscovery" images which suggest that McNeil's object is actually a variable nebula, similar to, but more extreme in its variations than, Hubble's Variable Nebula (see below). The discovery of this object in a bright state in early 2004 attracted a great deal of professional interest, and has helped to advance studies of a young (100,000 years old) star previously known only as an infrared source.

## *Some Challenging Reflection Nebulae*

### NGC 2261 Hubble's Variable Nebula
Monoceros    RA 06h 39.2m dec +08° 44′    mag. *c*.+10    Map 2

Young stars are often closely associated with the nebulosity from which they formed. A couple of good examples, albeit testing targets for small telescopes, can be found in the next Galactic spiral arm outward from our own, in the Taurus–Monoceros region. The stars that illuminate the dust in these nebulae have not yet quite settled into their stable main sequence state, and vary erratically in brightness over a magnitude or so. The prototype of this class of variable star (sometimes described as *nebular variables*) is T Tauri, associated with Hind's Variable Nebula (NGC 1554/1555, RA 04h 21.8m, dec +19° 32′) just north of the Hyades. T Tauri is probably less than 10 million years old, and currently varies between mag. +9.3 and +10.7. The associated nebula is fainter than 10th magnitude, and with a small angular span of 30″–40″ is a rather difficult prospect in typical amateur equipment.

Associated with the variable star R Mon (range mag. +11 to +14), Hubble's Variable Nebula (NGC 2261) is brighter and a more feasible target. Located just SE of the 5th-magnitude star 15 Mon (the second in a curved chain of four faint naked-eye stars trailing SW from the feet of Pollux in Gemini), this nebula is fan-shaped and about 2′ long, with the narrow tip pointing south. Usually around 10th magnitude, NGC 2261 is visible under good conditions in a 100 mm aperture telescope. At times it can be quite obvious, with a relatively high surface brightness.

Hubble's Variable Nebula shows changes in both brightness and outline on a timescale of weeks, as first noted by Hubble in 1916. Some of the changes in profile are a result of shadows cast by dark material between the nebula and R Mon. NGC 2261 lies 2600 light years away and was discovered by William Herschel in 1783.

### NGC 1435 Merope Nebula
Taurus    RA 03h 46.1m dec +25° 47′    Maps 2, 7

Long-exposure photographs and CCD images show the stars of the Pleiades (p.114) to be swathed in diaphanous, filamentary nebulosity. This was once thought to be material left over from the formation of the cluster, but although young at 76 million years, the Pleiades are old enough to have lost their nebulous cocoon. The dusty material draped around the Pleiades is in fact an interstellar cloud through which their motion is currently carrying them. Although prominent photographically, the Pleiades nebulosity is elusive for visual observers. This is partly because of its low surface brightness/contrast, but the glare from the bright Pleiades stars themselves also makes it hard to detect the nebula.

Some observers claim to have seen the nebulosity around the Pleiades in binoculars. Most, however, would recommend at least a small telescope and low magnification. The brightest section is NGC 1435, the Merope Nebula, named for the nearby mag. +4.1 star. Merope is at the SE corner of the Pleiades "mini dipper" (p.115). The nebulosity extends southward from Merope, with an overall extent of 30'.

The best way to look at NGC 1435 is to place Merope itself just outside the telescope's field of view – but beware of reflections or ghost images, produced by the telescope's optics, which may look deceptively like nebulosity. Instruments of 80 to 100 mm aperture *should* show NGC 1435 under good conditions; the nebula appears simply as a structureless haze in such telescopes. It is also known as Tempel's Nebula, after the German comet-hunter Wilhelm Tempel (1821–89), who discovered it in October 1859.

## Dark Nebulae

Unlike bright emission or reflection nebulae, dark clouds of interstellar material can reveal their presence only by blocking out the light from stars beyond. For this reason, the easiest dark nebulae to observe are those seen in regions where the background star density is high (close to the plane of the Milky Way) or in front of bright nebulae. Most prominent is the large (6.5° × 5°) Coalsack in the southern constellation Crux, seen as a great gap in the rich starfields. At a distance of 550 light years, the Coalsack (centered on RA 17h 52m, dec −63° 18') has been known since prehistoric times and is perhaps 50–60 light years across.

Other dark nebulae are less conspicuous, and most become apparent only on long-exposure images. Many of the best examples were listed by the revered American observer Edward Emerson Barnard, a pioneer of wide-field astronomical photography. From his photographs of the Milky Way, Barnard identified a number of dark nebulae, which he cataloged, and are now known by their "B" numbers – the visually elusive Horsehead Nebula in Orion is B33, for example. Some of the most prominent Barnard dark nebulae lie in the direction of the Galactic hub, and can be seen in binoculars under good conditions, particularly by observers at southerly latitudes. Indeed, low-power binocular views are often preferable to those in a larger instrument at high magnifications.

| B59 | | | |
|---|---|---|---|
| Ophiuchus | RA 17h 21.0m dec −27° 00' | Map 5 | Finder chart p.109 |

| B78 | | | |
|---|---|---|---|
| Ophiuchus | RA 17h 33.0m dec −26° 00' | Map 5 | Finder chart p.109 |

Seen against the rich Milky Way fields of Ophiuchus, east of and roughly level in declination with Antares and a couple of degrees south of

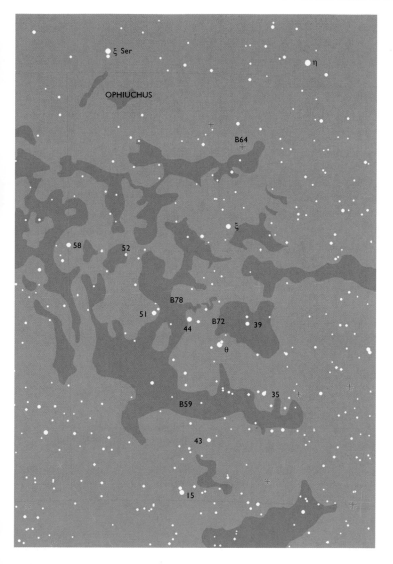

ξ Ser

η

OPHIUCHUS

B64

ξ

58          52

B78

51        B72     39
      44

      θ

                      35

B59

      43

      15

| 2 | 4 | Globular Clusters | Open Clusters |
| 6 | 8 | Dark Nebulae | |
| Magnitude of Stars | | | |

▲ The starfields of Ophiuchus, to the north of Scorpius and not far from the Galactic hub, abound with dark nebulae. The rough outlines of some of these are shown on the chart, which has an angular width of 12° and shows stars to limiting magnitude +8.5.

3rd-magnitude θ Oph, B59 and B78 form the stem and bowl respectively of a 7° long dark nebula popularly known as the Pipe. B59 lies roughly E–W.

### B72
Ophiuchus   RA 17h 23.5m dec −23° 28′   Map 5   Finder chart p.109

About 1.5° north of θ Oph, B72 is a region of dark nebulosity nicknamed the Snake.

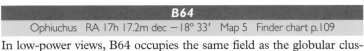
### B64
Ophiuchus   RA 17h 17.2m dec −18° 33′   Map 5   Finder chart p.109

In low-power views, B64 occupies the same field as the globular cluster M9 (p.86); it lies 30′ to the west of the globular. Dark material associated with the 20′-wide B64 probably dims our view of M9.

### B92
Sagittarius   RA 18h 16.9m dec −18° 02′   Map 5

Spanning 15′ × 9′, B92 is prominent against the rich bright detached section of Milky Way cataloged as M24, just north of the 4th-magnitude star μ Sgr.

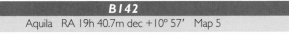
### B133
Aquila   RA 19h 06.1m dec −06° 50′   Map 5

Set against the bright Scutum Star Cloud, B133 covers an area of 10′ × 3′, 2° south of mag. +3.4 λ Aql and a couple of degrees east of the marvellous Wild Duck open cluster M11 (p.128).

### B142
Aquila   RA 19h 40.7m dec +10° 57′   Map 5

B142 lies just west of γ Aql in the Milky Way, and is perhaps the best-placed dark nebula for observers at northern latitudes.

# 6 · OPEN CLUSTERS

Open clusters are the products of the ongoing process of star formation in the H II regions and dark nebulae found in the spiral arms of our Galaxy. The density waves that create the spiral arms, or shock waves from supernova explosions, compress the nebular material and give rise to localized eddies. Dust grains adhere to one another, growing into larger clumps which are drawn to the center of the rotating, disk-shaped eddies under gravity, accreting more and more material until bodies of stellar mass, called protostars, are formed. Once a protostar has accreted sufficient mass, nuclear fusion is initiated in its core and it begins to shine.

As happened around our Sun 4.6 billion years ago, smaller, planetary bodies may also form around a star during its accretion. From studies of certain primitive meteorites – leftovers from planet formation – scientists have determined that formation of the Sun and planets took something like 30 million years, a short time in cosmological terms.

Most Galactic H II regions contain enough mass to spawn thousands of stars. The passage of a density wave through an H II region thus forms clusters of stars – stars seldom form in isolation. Over time, as it orbits the Galaxy, the members of a cluster will be subjected to the gravitational pull of other clusters and of giant molecular clouds in the spiral arms, and will gradually disperse into the disk population of stars. Most of the open star clusters we see are relatively young: those from the earliest history of the Galaxy have mostly long since scattered.

We can see star formation in progress in the heart of the Orion Nebula. In objects such as the Lagoon Nebula (M8) we see later stages in the process, where considerable amounts of gas and dust remain associated with a young star cluster. More aged clusters, such as Praesepe (M44), appear free of dust and gas.

The waves of star formation propagating around the Galaxy's spiral arms give rise to large-scale but (on cosmological timescales) short-lived structures called *OB associations*, among which the youngest open clusters are usually located. OB associations are so named for the massive, highly luminous stars of spectral types O and B which populate them: the blue supergiant Rigel in Orion is a good example of a B-type star. Many of the naked-eye stars in Perseus are part of an OB association, while the Scorpius–Centaurus OB association spans a wide expanse of the southern sky.

The ages of open clusters are estimated, as for globular clusters (Chapter 4), by plotting a Hertzsprung–Russell diagram for the member stars. The turn-off point from the main sequence to the giant branch is an indicator of age. Even massive stars take some time to evolve into red giants, and young clusters lack such objects. Several red giants are found in M41, for example, and this cluster is around 190 million years old.

Another clue to the age of an open cluster is given by the metallicity of their stars – their proportion of elements heavier than helium. Clusters formed relatively recently in nebulae rich in heavier elements synthesized by nuclear reactions in previous generations of stars will have a higher average metallicity than those dating from earlier epochs. Open star clusters belong to Baade's Population I, and are much younger than the metal-poor Population II stars in ancient globular clusters.

Hardly surprisingly, open clusters tend to be found mainly close to the plane of the Milky Way, which is of course where we look into neighboring spiral arms of the Galaxy. The closest open cluster is the Ursa Major Moving Cluster, to which five members of the familiar Big Dipper belong. These stars share a common motion through space, and are heading in the general direction of Sagittarius. They lie about 75 light years away.

▲ Two naked-eye open clusters are seen in this wide-field view of Taurus. At the left is the V of the Hyades, while the more compact Pleiades lie just right of upper center. The planet Jupiter is below the Pleiades with Saturn to its west (right) in this photograph taken by the author in February 2000.

Rather more obvious as open clusters to the naked eye are the Hyades and Pleiades in Taurus, and Melotte 111 in Coma Berenices. Each of these clusters is close enough for its member stars to be separated without optical aid. More distant open clusters, such as Praesepe, appear as hazy patches to the naked eye and require at least a pair of binoculars to be resolved into their individual stars; others may appear as "knots" against the Milky Way background. The more remote open clusters will, of course, be accessible only with a telescope.

Open clusters vary considerably in their appearance. Partly, these differences reflect their intrinsic nature: some are densely populated, others more sparse. Distance is also a factor, as discussed above – nearby clusters will tend to appear larger and more spread out, while those seen over large distances may show only as hazy, almost nebulous spots. Open clusters in rich Milky Way starfields may be hard to distinguish from their already crowded surroundings.

Open clusters come in a multitude of apparent shapes, too. Some are fairly circular, others drawn-out and elliptical, still others simply irregular collections of stars. A useful classification system for open clusters, devised by the Swiss-American astronomer Robert J. Trumpler (1886–1956), in given in Table 2. For example, Praesepe can be described as a Trumpler class IIm open cluster, while the rich Wild Duck Cluster M11 (NGC 6705) in Scutum is classed as Ir.

| TABLE 2: THE TRUMPLER CLASSIFICATION FOR OPEN CLUSTERS | |
|---|---|
| I | Detached stars with strong central condensation |
| II | Detached stars with little central condensation |
| II | Detached stars with no central condensation |
| IV | Unresolved concentration in an already rich starfield |
| Subdivisions | |
| p | Poor (less than 50 stars) |
| m | Medium (50–100 stars) |
| r | Rich (more than 100 stars) |

Open clusters are ideal objects for observation with binoculars and small telescopes. Many have loose scatterings of reasonably bright stars which render them more easily visible in light-polluted conditions than, say, diffuse nebulae or low-surface-brightness galaxies. As with other deep sky objects, it is worth viewing open clusters at a range of magnifications. A cluster which appears nebulous at low powers may be resolved into a rich haze of faint, separate stars in a higher-magnification eyepiece. It is interesting to compare clusters – the Puppis pair of M46 and M47, separated by only 1.5°, are quite different from each other, for example.

Sketching bright, loose open clusters is quite straightforward. Start by marking the relative positions of the brightest stars, then add in the fainter stars around them. The usual method is to use a hard pencil, drawing larger black dots on a white background to represent the brighter stars. For populous clusters (e.g. M11), drawing the relative positions of individual stars is well beyond the capability of most observers. In such cases, as with globular clusters, the relative stellar density can be represented by shading, with only the brighter field stars shown individually.

From the known diameter of the field of view (p.36), the observer can estimate the angular diameters of star clusters. Sometimes this is trickier than it might sound, for in many instances it can be difficult to decide where the cluster ends and the background/foreground stars begin.

Being particularly abundant along the length of the Milky Way, open clusters are popular targets from June to September, when the Cygnus–Scorpius–Sagittarius stretch is best presented, and during December to February, when the fainter span from Cassiopeia and Perseus, south past Orion and on to Puppis, Crux and Centaurus is on view in the evening sky, from one hemisphere or the other.

Open clusters are, of course, found in other galaxies, too, but typical amateur telescopes have little chance of resolving them. Mottling in the spiral arms of some face-on galaxies is indicative of the presence of star-forming H II regions, and of star clusters and OB associations.

## Selected Open Clusters

### M45 Pleiades
Taurus   RA 03h 47.0m dec +24° 07'   mag. +1.2   Maps 2, 7

Arguably the finest deep sky object visible to the naked eye, the tight grouping of the Pleiades in Taurus should be familiar to every observer. This is a cluster of quite young stars, still gravitationally connected 100 million years after their formation. The brightest cluster members are hot bluish stars of spectral type B. The Pleiades may contain around 500 stars in a volume of space 13 light years across.

To the naked eye, the Pleiades are an obvious compact cluster, 8° NW from the V of the Hyades and sometimes described as a spot on the Bull's shoulder. Most observers (the author included) can see six separate stars under dark sky conditions. Those with keen eyesight may be able to resolve and detect up to ten Pleiads, and some have even claimed to see as many as 14 stars here with the naked eye.

The Pleiades cluster is larger in angular extent than many people first think – it covers an area four times as wide as the Moon. And as it lies close to the ecliptic, it is occasionally occulted – obscured by – the Moon. These occultations are especially impressive when they occur in the evening sky and the Moon is a narrow crescent, its non-sunlit portion faintly visible by reflected earthshine. The stars' abrupt disappearance as the dark limb advances over them is spectacular in binoculars and small telescopes – and it takes several hours for the Moon to traverse the cluster, from west to east.

The best way to view the Pleiades is indisputably with a pair of binoculars. The cluster's 2° span fits well into the 5° field of 10 × 50s, allowing the observer to take in some of the hundreds of fainter stars scattered around the principal bright cluster members, which form a "mini-dipper" pattern – like the Big Dipper, but with only one star in its eastward-pointing handle.

At magnitude +2.9, the brightest Pleiades star is Alcyone ($\eta$ Tau), on the NE corner of the mini-dipper's bowl. Five of the other stars are brighter than 5th magnitude. I find that in a moonlit sky the difference in brightness between Alcyone and the other Pleiads becomes quite marked to the naked eye. Among the other bright Pleiades stars, Pleione (also given the variable star designation BU Tau), at the handle of the mini-dipper, is another hot star of spectral type B. Pleione rotates rapidly, and is apparently prone to "shell" episodes during which it throws off its outer layers, causing it to vary between mag. +4.8 and +5.5.

In binoculars and small telescopes at low powers, the blue color of the brighter Pleiades stars is quite pronounced. Drawing the myriad stars visible here is an exacting task: even binoculars reveal a couple of hundred down to 9th and 10th magnitude. Associated with the Pleiades is some faint reflection nebulosity, the brightest part of which is close to Merope, the mag. +4.1 southeastern star in the mini-dipper's bowl (NGC 1435; p.107). Most of the nebulosity is too faint for small amateur instruments, being best revealed in long exposure photographs and CCD images.

The Pleiades lie 415 light years away, well beyond the Hyades. Astronomers calculate that the stars in this currently compact cluster will be scattered by gravitational perturbations over the next 250 million years.

In ancient Egyptian times, the Pleiades' heliacal rising – the occasion each year when they rose in the dawn sky just ahead of the Sun – was a significant event, heralding the annual Nile floods and planting season. In northerly climes, their appearance in the late evening eastern sky during September is nowadays seen as ushering in the long dark nights of the prime observing season for North American and European amateur astronomers.

## Hyades
Taurus    RA 04h 27m dec +16° mag. +0.5    Maps 2, 7

Most of the stars in the V that marks the head of Taurus are members of an open cluster 150 light years away. The exception is mag. +0.8 Aldebaran (α Tau) – the Bull's baleful orange-red eye – which is a foreground object in the same line of sight, 68 light years distant. The Hyades are, indeed, the second-closest star cluster after the Ursa Major Moving Cluster.

With an angular size of 4° × 3°, the Hyades' main grouping is made up of 3rd- and 4th-magnitude stars, with a scattering of fainter members. Low-power binoculars offer the best view: a pair of 10 × 50s will just contain the whole V in the field. This core region of the cluster – outliers have been identified beyond the boundaries of Taurus! – has an actual size of about 15 light years.

The Hyades stars share a common motion through space, and are heading in the direction of a spot of sky to the east of Betelgeuse. They are very much older than the Pleiades (none of the hot, bright bluish stars found in the latter are evident), with an estimated age of 790 million years. In all, roughly 200 stars are thought to belong to the Hyades.

About 3° NE of Aldebaran is a fainter Taurus open cluster, NGC 1647 (RA 04h 46.0m, dec +19° 04′, mag. +6.4). Comprised of around 60 stars of mag. +9 in a circular area 25′ across, this is visible in binoculars on a good night.

## Mel 111
Coma Berenices    RA 12h 25m dec +26° mag. +1.8    Maps 3, 4

The main concentration of stars in Coma Berenices, a northward-pointing wedge about 4.5° long, east of Leo's hindquarters, certainly has all the appearances of a loose naked-eye cluster, its brightest members being of 4th to 5th magnitude. It was not until careful study by

◀ Many of the stars making up the "tresses" of Coma Berenices are members of a nearby cluster, Mel 111, photographed here by Nick Hewitt.

Trumpler in 1938, however, that the existence of a true cluster here was confirmed. Mel 111 (thus named in a catalog of objects published in 1915 by Philibert Jacques Melotte, 1880–1961) is a rather sparse cluster, with about 40 members; many of the stars in this region are chance alignments, but the genuine cluster members are identified by their common motion.

Mel 111 is easily seen with the naked eye on a good night, and certainly shows well in 7 × 50 or 10 × 50 binoculars. Its brightest star is of mag. +4.4, and there is a notable dearth of fainter members. Astronomers believe that Mel 111 is an aged (400 million years old) cluster on the verge of dispersing completely into the Galactic disk, having already lost its low-mass, fainter constituents to gravitational perturbations. It is the third-closest open cluster, at a distance of 288 light years.

## IC 2602 Southern Pleiades
Carina   RA 10h 43.2m dec −64° 24′   mag. +1.9   Map 8

Taking its popular name from a description by Lacaille, IC 2602 is a bright knot of stars in an area 50′ across. Also known as the θ Car cluster, from the 3rd-magnitude star at its center, IC 2602 lies at a distance of 479 light years.

## M44 NGC 2632 Praesepe
Cancer   RA 08h 40.1m dec +19° 59′   mag. +3.1   Map 3

Praesepe (Latin for "manger") in the heart of Cancer, midway between Castor and Pollux in Gemini and the Sickle of Leo, is an easy naked-eye object on any clear, dark night, appearing as a hazy patch slightly elongated N–S. Known since antiquity as a "little cloud," it was first resolved into stars by Galileo Galilei (1564–1642) in 1611 during his telescopic explorations. In binoculars, this is a very fine object – a scattering of stars of 6th magnitude and fainter, covering an area about 80′ (almost three Moon-widths) across.

Seen in 10 × 50 binoculars, M44 lives up to its alternative popular name of the Beehive, taken from its resemblance to a swarm of bees around a central quadrilateral asterism (I always think of this as a miniature version of Hercules' Keystone) made up by four of its most prominent members. The "Beehive" description first appeared in the *Cycle of Celestial Objects*, published in 1844 by Admiral William Henry Smyth (1788–1865).

Binoculars show more than 50 stars to mag. +9, and even a small telescope will show a great many more; the cluster may have as many as 200 members, and chance alignments with other, non-cluster stars give a total approaching some 350 stars in this area. In my 80 mm

◀ *Praesepe (M44) is a rich open cluster for binocular observers, with a prominent quadrilateral of brighter stars at its center, as shown in this image by Chris Cook.*

refractor, a magnification of ×20 is sufficient to provide a good view. At higher powers, parts of the cluster will lie outside the field of view.

M44 lies close to the ecliptic and can, like the Pleiades, be occulted by the Moon when circumstances are right. The planets Mars and Jupiter may be seen passing in front of the cluster in the course of their orbital motion; Jupiter was very close to the eastern edge in early 2003. Praesepe lies at a distance of 580 light years, and is a Trumpler class Ir cluster. Analysis of its stars gives M44 an age of 730 million years.

## M67 NGC 2682
Cancer   RA 08h 50.4m dec +11° 49'   mag. +6.9   Map 3

Cancer's second bright open cluster is found 8° SSE of Praesepe, 2° to the west of mag. +4.1 α Cnc. Much more distant than M44, this cluster is compact: while easily visible in 10 × 50 binoculars, M67 shows only as an oval haze, the long axis lying E–W, and spanning 30'. Binoculars will show a single (6th-magnitude) star on M67's eastern fringe, standing out from the background. In a small telescope at a medium power (e.g. ×40), this star again stands out, and the main body of the cluster is seen to comprise about a hundred stars, most mag. +7 and fainter. M67 is a Trumpler class IIr cluster.

Discovered by Johann Gottfried Köhler (1745–1801) in the late 1770s, M67 lies at a distance of 2700 light years, and is remarkably ancient, containing a number of well-evolved red giant stars. Given its estimated age of 3 billion years, this cluster has remained together very much longer than most: its position somewhat out of the Galaxy's disk has spared it the gravitational perturbations that scatter most clusters over cosmologically short timescales.

## Double Cluster
| Perseus | NGC 869 | RA 02h 19.0m dec +57° 09' | mag. +5.3 | Map 1 |
| Perseus | NGC 884 | RA 02h 22.4m dec +57° 07' | mag. +6.1 | Map 1 |

Located at the northern end of Perseus, near the border with Cassiopeia, the Double Cluster is a prominent haze, about a degree long and elongated E–W, visible to the naked eye. Surprisingly, although he included objects whose stellar composition is obvious – like the Pleiades – Messier did not list the Double Cluster in his catalog. Hipparchus did mention it as one of a number of "clouds" in his second-century BC star catalog. The two components of this fine object have separate entries in the NGC, and the Double Cluster is also known as the Sword Handle, and has the alternative designation of h and χ Per (NGC 869 is h, NGC 884 is χ).

While not resolved by the naked eye, the Double Cluster is seen as a pair of overlapping circular masses of faint stars, well contained in the field, in 10 × 50 binoculars; the stars range in brightness from 7th magnitude down to the detection limit of around mag. +10. NGC 869, the more westerly, has a slightly brighter catalog integrated magnitude, partly because its core region is more concentrated than that of NGC 884. In the binocular view, each cluster is a rich, fairly coarse scattering of a couple of hundred stars.

A small telescope will reveal many more faint stars within each component of the Double Cluster. A low magnification is preferable; each of these clusters has an apparent diameter of nearly 30', and at ×40 in my 80 mm refractor the two just fit into the 1.3° field. Higher powers will show only parts of the Double Cluster, but will help to emphasize the colors of the several prominent red giant stars here – indicators of the clusters' ages. Respectively, NGC 869 and NGC 884 are 5.6 million years and 3.2 million years old – comparatively young, but old enough for some of their most massive stars to have evolved into red giants. The difference in age between the two has been taken to suggest that they formed successively as a density wave passed

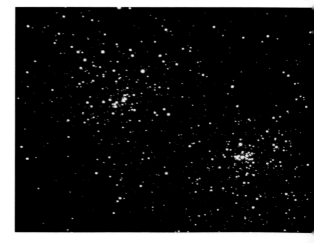

▶ The Double Cluster (NGC 869/884) in Perseus is a rich pair of overlapping objects, containing a number of prominent red giant stars, as can be seen in this photograph by Nick Hewitt.

through nebulosity in the Galaxy's Perseus spiral arm, the one beyond our own. NGC 869 lies 7100 light years away; NGC 884 is slightly more distant at 7400 light years.

The Double Cluster is an excellent object for northern observers during their winter – from North America and the British Isles, NGC 869 and 884 pass more or less overhead in late evening during December, and, viewed with binoculars from the comfort of a lounger, are a spectacular sight.

### M34 NGC 1039
Perseus   RA 02h 42.0m dec +42° 47'   mag. +5.2   Map 7

Easily found midway between Algol (β Per) and γ And, M34 is a sparse collection of about 40 mag. +6 to +8 stars, visible to the naked eye as a fuzzy spot from a dark location on a good, moonless night. Binoculars and finder telescopes perhaps offer the best views. M34 spans a diameter of about 35' (slightly larger than the Moon's apparent width). The cluster was discovered in August 1764 by Messier. It lies 1500 light years away and is estimated to be 100 million years old.

### NGC 1528
Perseus   RA 04h 15.4m dec +51° 14'   mag. +6.4   Maps 1, 2, 7

Found a couple of degrees ENE of mag. +4.3 λ Per, and about a degree NNW of 5th-magnitude b[1] Per, NGC 1528 is a contender for one of the best "escapees" from Messier's list of (low-power) comet-like diffuse objects. It is easily visible in 10 × 50 binoculars as an elongated, slightly curved 6th-magnitude hazy patch, aligned SSE–NNW; in binoculars, the long axis spans about 20'. The cluster is not quite resolved in 10 × 50 binoculars, but appears to have 7th-magnitude stars standing out from the general haze at either end.

In my 80 mm refractor, NGC 1528 is clearly resolved into stars and is a very fine cluster indeed. It contains about 40 stars to mag. +8/+9, with a line of mag. +7.5 to +8 stars marking the northern edge. The overall profile – in accordance with the binocular view – is a somewhat open C shape. NGC 1528 is of Trumpler class IIm – medium rich without a central condensation – and is a fine, if little-known, gem of the northern Milky Way.

## The Auriga Trio

### M38 NGC 1912
Auriga   RA 05h 28.7m dec +39° 50'   mag. +6.4   Map 2

### M36 NGC 1960
Auriga   RA 05h 36.1m dec +34° 08'   mag. +6.0   Map 2

▶ *M38 in Auriga is a rich collection of faint stars, well seen in small telescopes. This photograph by Nick Hewitt shows some of its "cruciform" star chains.*

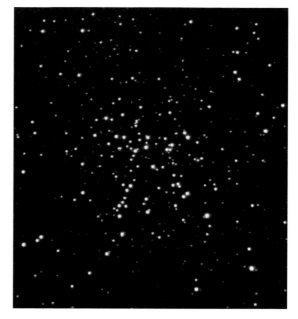

## M37 NGC 2099
Auriga   RA 05h 52.4m dec +32° 33′   mag. +5.6   Map 2

The northern Milky Way running through the west side of Auriga as it heads toward Gemini and Monoceros contains three fine Messier open clusters, each quite easily visible in binoculars. Although they appear pretty similar at first glance, there are many subtle differences. M38 and M36 were discovered by Guillaume Le Gentil (1725–92) in 1749; curiously, he missed M37, the brightest of the trio, which was discovered by Messier in 1764.

M38 is the most northerly of the three, found midway between θ and ι Aur (the broad side of the Auriga pentagon). It can also be found by taking a line from Capella to β Tau (Al Nath, marking the Bull's northern horn): M38 lies two-thirds of the way along this line from Capella. The cluster is rich (Trumpler class IIr), containing about a hundred stars in an area 20′ across. In 10 × 50 binoculars or a finder telescope, M38 is a hazy patch, brighter in its northern half. Telescopes as small as 60 mm aperture will resolve it into a swarm of tightly packed magnitude +8 and fainter stars. Many observers see a cruciform or T-shaped pattern in the stars here.

Lying 2.5° SE of M38, M36 is more compact, with an angular diameter of 12′. It is also a little less rich (Trumpler class Im), containing about 60 stars. As with its neighbor, binoculars don't quite resolve it, but M36 can be seen as a scattering of faint stars in any small telescope at a magnification of ×20 or more.

Further SE by 5°, M37 is also a hazy patch in binoculars, resolved in an 80 to 100 mm aperture telescope at ×40 into a 24′-wide circular mass of about 150 faint stars. It has a Trumpler class Ir – condensed and rich. M37 is the oldest of the three clusters, the presence of several well-evolved red giants indicating an age of 300 million years; M38 is 220 million years old, while M36 is a comparatively youthful 25 million years. In order of distance, M37 is 4400, M38 4200, and M36 4100 light years away.

Also worth a look in good conditions is NGC 1907 (RA 05h 28.0m, dec +35° 19′, mag. +8.2), an 8′-diameter cluster of about 30 stars just south of M38.

## M35 NGC 2168
Gemini   RA 06h 08.9m dec +24° 20′   mag. +5.1   Map 2

Visible to the naked eye under good conditions as an extended hazy spot just NW of 3rd-magnitude η Gem (Castor's "foot"), M35 is a rich, circular cluster containing around 150 stars in an area 28′ across – only a little smaller than the Moon's apparent diameter. The cluster is resolved in 10 × 50 binoculars, appearing as a haze of 8th- to 9th-magnitude stellar points. This impression is reinforced in a small telescope, while at ×100 in my 102 mm refractor, M35 fills the 31′ field with a coarse scattering of faint, white stars. The cluster has a Trumpler class IIIr.

M35 has an estimated age of 100 million years. It is similar in appearance to the Auriga Trio of clusters not far to its north, but rather closer to us, at a distance of 2800 light years. Its actual diameter is about 24 light years.

## NGC 2158
Gemini   RA 06h 07.5m dec +23° 18′   mag. +8.6   Map 2

Careful examination of M35 with a 150 mm aperture telescope will reveal what appears to be a more condensed region in the cluster's SW edge. This is in fact another, more remote cluster in the same line of sight: NGC 2158 lies 16,000 light years away. Covering an angular diameter of 5′, NGC 2158 was discovered by William Herschel in 1784. The cluster is very ancient, with an estimated age of 1.05 billion years.

## M50 NGC 2323
Monoceros   RA 07h 03.2m dec −08° 20′   mag. +5.9   Map 2

The span of the Milky Way east of Orion, heading south from Gemini through Monoceros toward Puppis, best presented during December and January, is much dimmer than the Cygnus–Sagittarius region which is so impressive on June and July nights. It is, however, graced by many excellent open clusters. Among these is M50, found on the line between Sirius and Procyon, about a third of the distance between the two, 8° NNE of the former. M50 is a reasonable binocular object, just resolved into a scattering of faint stars in 10 × 50s. Small telescopes show it well, as a collection of about a hundred stars to mag. +9 in an area 15′–20′ across; many of the fainter members become apparent only in averted vision. Discovered by Messier in April 1772, M50 is about 3000 light years away, and 18 light years in actual diameter. It has a Trumpler classification of IIm – moderately rich and not particularly condensed.

## NGC 2244

Monoceros   RA 06h 32.4m dec +04° 52′   mag. +4.8   Map 2

The cluster of young stars (perhaps only half a million years old) at the heart of the Rosette Nebula (p.104) is an attractive object in its own right for small telescopes. Easily found at the apex of a flat triangle it forms with 13 Mon and ε Mon (the last two in a curved chain of four faint naked-eye stars running SW from the foot of Pollux in Gemini), the cluster is seen as a rectangle of 6th- and 7th-magnitude stars with its longer axis lying SSW–NNE, fitting neatly into the field of view at ×30. The cluster is just visible to the naked eye as an unresolved spot, and is split into individual stars in 10 × 50 binoculars. A total of 30 stars are located in its 24′ span.

Among Monoceros' other clusters, NGC 2264 (RA 06h 41.1m, dec +09° 53′, mag. +3.9) is worth seeking out, and includes as its brightest member the 5th-magnitude 15 Mon – second in the chain of faint stars SW of Gemini which serve as a guide for NGC 2244. NGC 2264 is nicknamed the Christmas Tree cluster from its 20′-long westward-pointing V of about 20 stars to mag. +8: 15 Mon is the "pot" in which the tree stands. The cluster is an easy object in small telescopes. This part of the Monoceros Milky Way is also home to the Cone Nebula, a dark notch in some faint emission nebulosity accessible only to the largest amateur telescopes.

## M41 NGC 2287

Canis Major   RA 06h 47.0m dec −20° 44′   mag. +4.5   Map 2

Easily located, 4° more or less due south of Sirius, M41 appears in small telescopes as a rather loose scattering of between 30 and 50 stars ranging from mag. +7 to +10 in a diameter of 38′. From a dark location it can be seen as a faint naked-eye spot, and in binoculars the cluster is obvious and attractive. The best views are to be had either in binoculars or in a small telescope at low powers; M41 fills the field in a 110 mm reflector at ×31, for example. The cluster's outstanding feature is a markedly red mag. +6.9 star near its relatively empty apparent center. Larger telescopes reveal many faint stars, suggesting an overall membership of a hundred or so. M41 is a Trumpler class Ir cluster and lies 2300 light years away. The presence of several evolved red/orange giant stars suggests that its age is about 190 million years.

## NGC 2362 Tau Canis Majoris Cluster
Canis Major   RA 07h 18.8m dec −26° 47′   mag. +4.1   Map 2

Rather more compact than its neighbor M41, NGC 2362 has a diameter of about 8′ and contains about 40 mostly rather faint (mag. +10 to +11) stars. These surround the intrinsically highly luminous magnitude +4.4 O-type star τ CMa, which may be a cluster member, and makes this an easy object to locate, some 9° SSE of Sirius. The view in a small telescope is perhaps rather less impressive than the integrated magnitude might lead one to expect. NGC 2362 is estimated to be less than a million years old, and lies at a distance of 4600 light years.

## M47 NGC 2422
Puppis   RA 07h 36.6m dec −14° 30′   mag. +4.4   Maps 2, 3

## M46 NGC 2437
Puppis   RA 07h 41.8m dec −14° 49′   mag. +6.1   Maps 2, 3

On more or less the same line of declination as Sirius, 15° to its east – and therefore reaching the meridian about an hour after Sirius – the Puppis pair of M46 and M47 show an interesting contrast in appearance. The two can be seen easily in binoculars, about 1.5° apart, with M47 being the more westerly.

M47 can be seen as a small, tight grouping of naked-eye stars on any reasonable night, resolving into individual components in 7 × 50 or 10 × 50 binoculars: about a dozen stars of mag. +5.5 to +6.5 are contained in an area 29′ across. In a telescope, they appear rather scattered, and the cluster is rather sparse (Trumpler class IIIm): a background of fainter components takes M47's overall stellar content to about 50.

M46 is beyond naked-eye reach, but is visible in binoculars as a slightly oval haze similar in size to M47, showing only hints of individual stars in 10 × 50s. In a small telescope, the haze dissolves into a rich background of fairly evenly bright 9th-magnitude stars, revealing this as a well-populated cluster (Trumpler class IIr). A 100–150 mm aperture telescope can show as many as 150 stars, together with the faint, foreground planetary nebula NGC 2438 (p.158), 7′ north from the cluster's center: at magnifications of ×100 and higher, the effect is pleasingly three-dimensional.

M47 is 1600 light years away and has a estimated age of 78 million years: M46 is more remote and aged, at 5600 light years and 300 million years old.

## M93 NGC 2447
Puppis   RA 07h 44.6m dec −23° 52′   mag. +6.2   Maps 2, 3

About 9° due south of the M46/M47 pair, a degree NW of mag. +3.4 ξ Pup, M93 is one of the more southerly Messier objects, but is still readily visible from UK and US latitudes. M93 is seen in 10 × 50 binoculars as a westward-pointing arrowhead of stars. In my 80 mm refractor at ×20, it resolves easily into a knot of at least 30 stars of 8th magnitude and fainter, a line of six brighter members marking its southern edge. Larger telescopes show up to 80 stars in this Trumpler class Ir cluster. M93 has an angular span of 22′. It lies 3600 light years away and has an estimated age of 100 million years.

Also in southern Puppis, 2.5° NW of the mag. +2.3 star ζ Pup, is NGC 2477 (RA 07h 52.3m, dec −38° 23′, mag. +5.8), a rich haze of about 150 stars in an area just under 30′ across. This and the nearby NGC 2451 (RA 07h 45.4m, dec −37° 58′, mag. +2.8) are best seen from southerly latitudes, and are out of reach for observers in Northwest Europe. NGC 2451 is a 45′-diameter scattering of 40 mainly faint stars dominated by mag. +3.6 c Pup.

## M48 NGC 2458
Hydra   RA 08h 13.8m dec −05° 48′   mag. +5.8   Map 3

Discovered by Messier in 1771, M48 is a fairly loose cluster in a rather empty part of the sky 20° due east of 2nd-magnitude Alphard (α Hya)

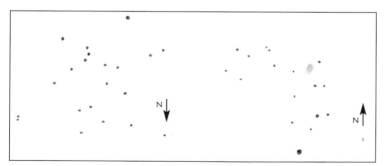

▲ Two sketches of M48 in Hydra, both made by the author in 1980. The left-hand view represents the cluster as seen in a small (114 mm aperture) Newtonian reflector at ×31. On the right is the wider-field view in 10 × 50 binoculars, with which the cluster is barely resolved.

and 10° SW of the circlet of faint stars that form the head of Hydra. Binoculars show M48 as a partly resolved haze, elongated SE–NW, which clearly separates into a collection of around 40 stars of mag. +7 to +9 in an area 40′ across. The cluster's central region is dominated by three yellow giant stars, giving the impression of a condensation. With an estimated age of 300 million years, M48 has a Trumpler class Im, and is about 2000 light years away.

## IC 2391
Vela   RA 08h 40.2m dec −53° 04′   mag. +2.5   Maps 3, 8

A southern object known since antiquity (it is mentioned in al-Sufi's star catalog of AD 964), IC 2391 is a bright cluster of about 30 stars scattered over 50′, and visible to the naked eye. Its loose nature (Trumpler class IIp) makes this an object best viewed in binoculars. The cluster is quite young (36 million years old) and lies 580 light years away.

## NGC 4755 Jewel Box
Crux   RA 15h 53.6m dec −60° 20′   mag. +4.2   Map 8

Taking its popular name from John Herschel's description – made during his expedition to South Africa in the 1830s (on which he surveyed the southern sky and observed Halley's Comet), NGC 4755 is a cluster of bright young stars near the Coalsack in Crux. Also known as the Kappa Crucis Cluster from the 6th-magnitude supergiant star at its center, NGC 4755 contains many strongly colored blue, red and yellow stars among its 50 or so members. It covers an area 10′ across and is a good object for binoculars and small telescopes. NGC 4755 lies 7600 light years away and has an estimated age of 7.1 million years. It was first described by Lacaille, following his observations in 1751–53.

## IC 4665
Ophiuchus   17h 46m dec +05° 43′   mag. +4.2   Map 5

Best known for its fine globular clusters, Ophiuchus also hosts a good open cluster for binocular and low-power telescopic viewing. Lying a degree NNW of mag. +2.8 β Oph, IC 4665 is a scattered cluster (Trumpler class IIIp) of 30 or so 7th- and 8th-magnitude white stars in an area 45′ across. Its most distinctive feature is a circlet of stars on the western side. IC 4665 is quite young (36 million years old) and lies 1400 light years away.

## M6 NGC 6405 Butterfly Cluster
Scorpius   RA 17h 40.1m dec −32° 13′   mag. +4.2   Map 5

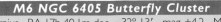

## M7 NGC 6475
Scorpius   RA 17h 53.9m dec −34° 49′   mag. +3.3   Map 5

This pair of bright open clusters close to Scorpius' Sting stars is tanta-lizingly positioned for observers in the British Isles, clearing the horizon by little more than 7° and 5° in southern England. Those in the United States fare a bit better, but the best views of M6 and M7 are to be had from more southerly latitudes, where both are naked-eye objects.

M6 is a scattered cluster of about a hundred faint stars (Trumpler class IIIp) spread over an area 25′ across. It gets its nickname of the Butterfly Cluster from the way its stars seem to form two "wings." Seen low in UK skies with 10 × 50 binoculars, it is an E–W elongated oval, partly resolved to reveal a few stars standing out from a hazy back-ground. M6 is 1600 light years away and 12 light years in actual size.

The most southerly of the Messier objects, M7 was known in antiq-uity (as, in all probability, was M6). M7 contains about 80 stars brighter than 10th-magnitude in an 80′ span; long-exposure images reveal asso-ciated nebulosity. In my restricted view from southern England, 10 × 50 binoculars show M7 to be dominated by about ten 5th- to 6th-magnitude bright stars, with a scattering of further barely visible mem-bers in the background. M7 is a Trumpler class Im cluster. It lies 800 light years away (and is therefore unconnected to M6); it has an actual diameter of about 20 light years and an age of 220 million years.

## M24 Small Sagittarius Star Cloud
Sagittarius   RA 18h 16.5m dec −18° 50′   mag. +4.6   Map 5

M24 is easily visible to the naked eye as a detached piece of the Milky Way, NE of the Teapot, and about 3° north of the 4th-magnitude star μ Sgr. The oval cloud is 1.5° along its longer dimension, which lies NE–SW, while the shorter axis spans about 45′. Small (10 × 50) binoculars show a rich mass of stars here in a very crowded field; in 25 × 100 binoculars, this object is awesome – a field-filling scattering of hundreds of stars with an almost three-dimensional appearance.

The appearance of this as a cloud is partly a line-of-sight effect: a gap in the dense gas and dust in the direction of the Galactic center allows us to see through to the rich starfields in a far distant interior spiral arm. The dark nebula B92 (p.110) is seen in contrast against M24's NW edge, while the Swan Nebula (M17; p.102) lies just to the NE.

## M25 IC 4725
Sagittarius   RA 18h 31.6m dec −19° 15′   mag. +4.6   Map 5

About 4° due east of M24, M25 is a reasonable cluster in 10 × 50 binoculars, appearing as a knot of about 10–12 mag. +5 and fainter stars in an area 40′ across. Small telescopes show a further 20 or so stars, but although it is quite prominent in binoculars, this is a relatively sparse (Trumpler class Ip) object. It lies 2000 light years away and has an estimated age of 90 million years.

## M11 NGC 6705 Wild Duck
Scutum   RA 18h 51.1m dec −06° 16′   mag. +5.8   Map 5

Found in the rich Milky Way fields of the Scutum Star Cloud, M11 is among the finest open clusters. A good guide to location is the line of stars pointing SE from λ and 12 Aql to the mag. +4.8 η Sct; M11 lies just west of η, and to the east of a parallelogram of 5th- to 7th-magnitude stars.

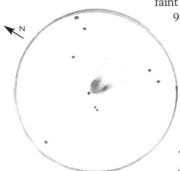

In 10 × 50 binoculars the cluster is a prominent, hazy unresolved circular spot. Small telescopes resolve this into a swarm of faint stars, all more or less equally bright (about 9th magnitude) in a fairly open V-shape pointing eastward. A brighter, mag. +7, star lies just off M11's eastern tip. In my 80 mm refractor at ×40, the cluster fills almost a fifth of the field and is simply magnificent – I would rank the rich M11 as my favorite open cluster. At least a hundred faint stars are seen here.

M11 was discovered by Kirch in 1681. The cluster's popular name comes from a description by Admiral Smyth in his *Cycle of Celestial Objects*, where he likens the cluster's outline to ducks in flight. The Wild Duck Cluster lies 6000 light years away and has an estimated age of 250 million years.

▲ *The author's sketch of M11, as seen in an 80 mm f/5 refractor at ×40. The individual stars in this rich cluster are too numerous and closely packed to draw. There is a brighter star at its eastern tip.*

A couple of degrees SW of M11 is Scutum's other Messier cluster, M26 (NGC 6694, RA 18h 45.2m, dec −09° 24′, mag. +8.0). It is rather less eye-catching than its neighbor, comprising 40 stars in a 14′ area.

## M52 NGC 7654
Cassiopeia   RA 23h 24.2m dec +61° 35′   mag. +6.3   Map 1

Cassiopeia lies in a rich part of the northern Milky Way and contains a number of good open clusters for binoculars and small telescopes. Brightest of these is M52, discovered in September 1774 by Messier. M52 is easy to locate: following the line from α to β Cas (the westernmost two of the W) NW by a little more than the 5° distance between them brings the observer to the cluster. In binoculars it is a

hazy, partly resolved circular patch about 12′ across. In any small telescope it will resolve into a mass of stars – reminiscent of the Auriga
Messier clusters – with more of a V-shaped outline. Most of the 150
or so cluster members are around 9th magnitude, and a brighter
(mag. +7) star lies close to M52's SW edge.

M52 is a concentrated Trumpler class Ir object. It is 3900 light years
away, giving an actual diameter of about 20 light years. Its age is estimated at 35 million years.

Also well worth a look in this part of Cassiopeia is NGC 7789 (RA
23h 57.0m, dec +56° 44′, mag. +6.7), just south of the 5th-magnitude
variable star ρ Cas and a couple of degrees from β at the western end of
the W. Viewed in binoculars from a dark site, this is a prominent hazy
circular patch 15′ across. In a telescope it resolves into a crowd of a hundred or so 9th-magnitude stars, and is a Trumpler class IIr cluster.

## M103 NGC 581
Cassiopeia   RA 01h 33.2m dec +60° 52′   mag. +7.4   Map 1

About a degree NW of δ Cas (the fourth star eastward in the W),
M103 is an attractive, easily located cluster of 7th- and 8th-magnitude
stars, visible in binoculars. A small telescope improves the resolution.
In a 114 mm reflector at ×31, I see this as a straggling pair of back-to-
back L shapes, the long arm consisting of two chains of stars aligned
SE–NW and extending for about 5′. M103 contains about 30 stars and
is given a Trumpler class IIIp – rather sparse. Discovered by Méchain
in 1781, M103 lies 9200 light years away and has a true diameter of 15
light years.

## NGC 457
Cassiopeia   RA 01h 19.1m dec +58° 20′   mag. +6.4   Map 1

Sometimes called the Phi Cassiopeiae Cluster, for the mag. +5.0 star
which appears to be its principal luminary, NGC 457 is an attractive
binocular object a couple of degrees SW of δ Cas. In 10 × 50s, NGC
457 appears on a good, dark night as a hazy oval, perhaps 15′ in diameter. Telescopically, this is a rich cluster of
faint stars (Trumpler class Ir) with a total of
about 60 members. NGC 457 lies 9300 light
years away.

In addition to NGC 457 and M103, the
area close to δ Cas contains several other fine

▶ NGC 457 is sometimes
called the "ET" cluster
for the brighter pair of

stars ("eyes") seen at
top in this photograph
by Nick Hewitt.

open clusters: NGC 663 (RA 01h 46.0m, dec +61° 15', mag. +7.1), NGC 654 (RA 01h 44.1m, dec +61° 53', mag. +6.5) and NGC 659 (RA 01h 44.2m, dec +60° 42', mag. +7.9).

## M39 NGC 7092
Cygnus   RA 21h 32.2m dec +48° 26'   mag. +5.0   Maps 1, 6

The rich northern Milky Way in Cygnus contains many faint open clusters, listed in the NGC; many of them appear as little more than enhancements of the already star-crowded background and are hard to pick out. Even the brightest of the Cygnus clusters is somewhat swamped by its stellar surroundings: M39, 9° ENE of Deneb, is a rather disappointing telescopic object but stands out reasonably well in $10 \times 50$ binoculars. The closest naked-eye star to M39 is the 4th-magnitude $\pi^2$ Cyg, a couple of degrees ENE of the cluster. M39 is a very loose collection of white stars, 30' across, the brightest of which is of mag. +6.5. M39 is a Trumpler class IIIp cluster, 800 light years away and 7 light years across.

## M29 NGC 6913
Cygnus   RA 20h 23.9m dec +38° 32'   mag. +6.6   Map 6

Discovered by Messier in 1764, M29 is reasonably compact and recognizable as a cluster, just under 2° due south of γ Cyg (the central star of the Northern Cross), containing about 20 stars of mag. +7 to +9 in a southward-pointing triangle with sides of about 7'. Lying at the considerable distance of 7000 light years, M29 is apparently dimmed by dust in the plane of the Milky Way.

## NGC 752
Andromeda   RA 01h 57.8m dec +37° 41'   mag. +5.7   Map 7

Located 4.5° SSW of the fine double star Almach (γ And; p.140), NGC 752 is a good binocular object, covering a large area 50' across. Its brightest stars are around 9th magnitude, making this a barely resolvable haze in $10 \times 50$ binoculars. A couple of mag. +6.5 stars lie to its south, and a triangle of 8th-magnitude stars to the north. The 60 or so stars that belong to this Trumpler class IIIm cluster are better resolved in a small telescope, but NGC 752's large angular size makes it an object best appreciated in wide-field views.

## Asterisms

Several apparent stellar groupings in the sky are the results of simple coincidence: stars that appear to be near one another can turn out to be at greatly differing distances and merely to lie on much the same line of sight. In some cases, several stars may line up together in a small angular area, producing the illusion of a cluster. While they are not genuine clusters, some of these compact groupings – called *asterisms* – make attractive observational targets. The term is also used for larger-scale celestial groupings, such as the seven stars of the Big Dipper or Plough (a subset of Ursa Major, and not a constellation in its own right); in fact five of the Big Dipper stars are members of the Ursa Major Moving Cluster (p.112).

## M40

Ursa Major   RA 12h 22.4m dec +58° 05′   Map 1

Two asterisms found their way into Messier's catalog. The first, M40, is 1.5° NE of δ UMa (the faintest of the Big Dipper stars) and consists, simply, of a pair of white stars of mag. +9.0 and +9.3. Separated by 49″, these can look nebulous on a casual inspection at a low power in a small telescope, but any modern "spotter" instrument of reasonable quality will resolve them. The stars may be a physically connected double system, as they lie at a distance of 510 light years. Messier was aware that this was a pairing of similarly bright stars when he added M40 to the first version of his catalog in 1771: the German celestial cartographer Johannes Hevelius (1611–87) had established the object's true nature over a century earlier in 1660. Most observers seek this out simply to complete their M-object checklist: it is not particularly distinguished!

## M73 NGC 6994

Aquarius   RA 20h 59.0m dec −12° 58′   Map 6

The second asterism in Messier's catalog lies close to the 9th-magnitude globular cluster M72 (p.83). M72 and M73 are at almost the same declination, so if the field is centered on M72 and the telescope left static for $5\frac{1}{2}$ minutes (the difference in RA between the two objects), M73 will be carried into the field by the Earth's rotation. M73 comprises four evenly matched 11th-magnitude stars in a Y formation about 3′ across, best seen in a small telescope at a medium power. The southernmost star of the four is marginally the brightest. Messier added the asterism to his list in October 1780. Unlike M40, M73 was afforded a listing in the 1888 *New General Catalogue*.

## Cr 399 Coathanger

Vulpecula   RA 19h 25.4m dec +20° 11′   Map 5

Visible to the naked eye as a partly resolved knot of faint stars 4° NW of the triangle of stars marking the tail of Sagitta's arrow, Cr 399 is perhaps the brightest and most attractive of binocular asterisms, and one which many observers will have come across while sweeping the star-

◀ The Coathanger in Vulpecula, just north of Sagitta's tail, is an attractive binocular asterism. It is seen here in a photograph by Nick Hewitt.

rich Milky Way fields between Aquila and Cygnus. The asterism was listed in a 1931 catalog of clusters compiled by the Swedish astronomer Per Collinder (1890–1975). Measurements of the stars' distances give values ranging from 280 to 1140 light years, so despite appearances this is not a true cluster.

Only a little smaller in angular extent than the Pleiades, Cr 399 takes its popular name of the Coathanger from its appearance as a 1.5°-long bar of six 5th to 6th-magnitude stars in a roughly E–W line, with a "hook" formed by a further four 5th to 6th-magnitude stars stretching 20′ to the south of its midpoint. The object is best viewed in the low-power field of a pair of binoculars. American observers sometimes refer to Cr 399 as Brocchi's Cluster, commemorating the celestial cartographer Dalmiro Francis Brocchi (1871–1955), who mapped it in the 1920s.

## Kemble's Cascade
### Camelopardalis   RA 04h 10m dec +63°

Named for the Canadian amateur astronomer Fr Lucien Kemble (1922–99), who drew attention to it, this asterism is a real treat for observers at northern latitudes, enlivening the otherwise dim and sparse constellation of Camelopardalis to the east of Perseus and north of Auriga. Kemble first drew attention to the asterism during the 1980s. Kemble's Cascade is an almost straight line of fairly faint (mag. +7 to +8) stars running SE–NW for a distance of about 2.5°.

The asterism is well contained in the field of a pair of 10 × 50 binoculars: in the low-power view, a 6th-magnitude star near the midpoint of the line becomes obvious. None of the stars making up the line – I have counted at least 17 – shows any pronounced visual color: the most remarkable features of Kemble's Cascade are its straightness and the evenly matched brightness of the component stars. This impression is reinforced in a medium-power telescopic view: at ×40 in a wide-view refractor, for instance, the stars just keep coming along the line for field after field!

Near the SW end of Kemble's Cascade is a genuine cluster, NGC 1502 (RA 04h 07.7m, dec +62° 20′, mag +5.7), triangular in outline and containing stars of 7th magnitude and fainter. The galaxy IC 342 (p.57) is also in the vicinity. Perhaps the best Camelopardalis cluster for small instruments is Stock 23 (RA 03h 16.3m, dec +60° 02′, mag. +8.4), about 14° west of Kemble's Cascade, and 15° east from the Double Cluster in Perseus. Stock 23 shows a tight grouping of four 6th-magnitude stars with a background of 20–25 fainter members.

▶ Kemble's Cascade, seen here in a photograph by Nick Hewitt, is a remarkably straight run of evenly matched stars ending at the genuine cluster NGC 1502 in Camelopardalis.

# 7 · DOUBLE STARS

Stars are born in nebulae in the spiral arms of galaxies, forming in clusters which gradually disperse under the influence of gravitational perturbations. Some stars, however, are sufficiently closely bound gravitationally to stay together as a pair, a *binary star* – or sometimes a multiple system with three, four or more components – after the rest of the cluster in which they were born has scattered. Binary or multiple systems may even be more common than solitary stars: our Sun may be unusual in lacking a stellar companion.

Binary stars revolve around a shared center of gravity, and can have orbital periods ranging from a matter of hours in close, tight binary systems to hundreds or thousands of years for widely separated pairs. Very close binaries are beyond the resolving power of the most powerful telescopes; their double nature can be inferred only from periodic shifts in their spectral lines, and they are therefore known as *spectroscopic binaries*.

Some very close binaries are progenitors of eruptive variable stars, in which matter is transferred from the outer atmosphere of a distended giant star onto the surface of a dwarf companion, leading to "nuclear runaway." At its most extreme, such runaway can trigger a Type I supernova event resulting in the progenitor's destruction. So-called classical novae are slightly less extreme, resulting in ejection of material but survival of the giant and dwarf, with the system showing repeated outbursts separated by long interludes of quiescence over tens of thousands of years. Dwarf novae and cataclysmic variables (SS Cygni and U Geminorum, for example) undergo smaller-scale, more frequent outbursts.

Still closely bound, but perhaps less likely to be undergoing mass transfer, are systems such as Algol, the famous eclipsing binary in Perseus. In eclipsing binaries, the orbital plane of a fainter companion around the brighter member of the system lies in our line of sight, and when one periodically passes in front of the other there is a dip in the overall brightness.

From the deep sky observer's point of view, the most interesting double stars are those which appear far enough apart – with sufficiently large *separations* – for the component stars to be distinguished. Many are true binary systems, also known as physical doubles, but some are so-called *optical doubles*, whose proximity in the sky is simply a coincidental, line-of-sight effect. The ability of a telescope or binoculars to separate stellar point sources on the sky depends on its aperture. This is measured in terms of the so-called Dawes limit, given by $116/d$, where $d$ is the telescope aperture in millimeters, as indicated in Table 3.

| TABLE 3: TELESCOPE RESOLVING POWER | |
|---|---|
| Telescope Aperture (mm) | Closest double star separation resolved (arcseconds, ") |
| 50 | 2.3 |
| 60 | 1.9 |
| 70 | 1.7 |
| 80 | 1.5 |
| 100 | 1.2 |
| 125 | 0.9 |
| 150 | 0.8 |

An interesting observational challenge is to see whether you can push your telescope to the theoretical limit of resolution. This is perhaps easier to do when observing doubles in which the stars have a similar magnitude. In some doubles (mag. −1.5 Sirius and its mag. +8.5 white dwarf companion is an obvious example), a bright primary can overwhelm the light of a faint secondary. The resolution of particularly close pairs will, of course, demand a high magnification, but the observer should also bear in mind the rule of thumb that a telescope's maximum useful magnification is typically about twice the aperture in millimeters. Unsteady seeing conditions will make it harder to separate close pairs.

It is usual to designate the primary (normally brighter) star A, its companion B (and any additional fainter companions C, D, and so on).

Serious double star specialists may spend much of their observing time making detailed measurements of the separation and position angle (p.20) of the components. In some systems the orbital period is short enough for the position angle of the secondary relative to the primary to change noticeably over the course of a few years. Just as the planets move around the Sun in elliptical orbits, the orbit of a secondary around the primary in a double star system is an ellipse, and in the more eccentric doubles, the change in separation between closest approach (called periastron) and greatest distance (apastron) can be substantial. Measurements of separation and position angle call for expensive, specialized micrometer equipment.

There are many, however, who simply observe double and multiple stars for the enjoyment it brings. Like many observers, I find the most aesthetically appealing double stars to be those which show a pronounced color contrast between primary and secondary (e.g. Gamma Andromedae) or in which the components are perfectly matched in brightness and color (e.g. Gamma Arietis). Colors can sometimes be enhanced by slightly defocusing the telescope. As "recreational" objects, double stars have the added advantage of being observable under bright moonlight, or the twilight of summer at high northern latitudes – in some respects, color contrast doubles actually benefit from the lighter sky background at such times. Also, double stars are one of the few types of astronomical object that can be observed reasonably well under light-polluted suburban skies.

A star's color is an indicator of its physical condition. The long-established sequence of spectral types – O, B, A, F, G, K, M – runs from hot, white O-type stars via yellowish, Sun-like type G stars to cooler orange K stars and red M stars. Some of these types are represented among the double stars accessible to amateur astronomers' telescopes – the primary of the Alpha Herculis system, for example, is a red M-type star.

The study of double stars was of particular interest to nineteenth-century astronomers, and many of their descriptions make interesting reading. The components were frequently ascribed such tints as topaz, amethyst, emerald, lilac and aquamarine. The apparent colors in many double stars are, of course, subjective. For instance, I see Albireo's primary as orange and its secondary as green, while others record the secondary as blue or turquoise. The selection below offers some relatively easy targets (some can be resolved even in $10 \times 50$ binoculars) and a few which are more testing. Observers are invited to form their own impressions! These stars represent but a fraction of the known double and multiple systems in our Galactic neighborhood: tens of thousands are cataloged, and observing them all would be a lifetime's work for the truly dedicated enthusiast.

## Selected Double and Multiple Stars

### Wide Doubles

| ζ UMa Mizar |
|---|
| RA 13h 23.9m dec +54° 56′   Maps 1, 4 |

As the middle of the three stars making up the handle of the Big Dipper, Mizar (mag. +2.3) and its mag. +4.0 companion, Alcor, is the best known naked-eye double. The pair's wide separation of 11.8′ is over a third the Moon's apparent diameter.

Inspection with any small telescope at a magnification of ×40 or more reveals that Mizar is itself a double, with white components of mag. +2.3 and +4.0 separated by 14.4″ in PA 152°. These have each been identified as spectroscopic binaries, as has Alcor. Alcor is gravitationally bound to Mizar, making this a sextuple system. Mizar and Alcor are part of the Ursa Major Moving Cluster of stars, 75 light years away.

► *Mizar (lower right) and Alcor form a wide naked eye double in the Big Dipper's handle. Telescopic examination shows Mizar itself to be a close double star, as seen in this photograph by Nick Hewitt.*

## α Cen Rigil Kentaurus
### RA 14h 39.6m dec −60° 50′   Map 8

A southern-hemisphere binary system with a period of 80 years, α Cen lies at a distance of 4.4 light years. The primary is a yellowish Sun-like G-type star of mag. 0.0, with a 14″ distant, more orange K-class secondary of mag. +1.5. The most celebrated member of this system is the third star, Proxima Centauri, a faint (mag. +11) red dwarf over 2° from the primary at a PA of 088°. At its distance of 4.2 light years, Proxima Centauri has the distinction of being the closest star beyond the Sun.

## o² Eri
### RA 04h 15.3m dec −07° 39′   Maps 2, 7

Relatively undistinguished to the naked eye as a 4th-magnitude star about 15° west of Rigel, o² Eri is interesting both historically and in terms of its membership. The primary is a slightly orange K-type star of mag. +4.4, with a mag. +9.5 companion a wide 83″ away in PA 104°. This companion star is the easiest white dwarf star to observe, being visible in an 80 mm refractor in good conditions. The white dwarf has a tricky, mag. +11, red dwarf companion at an angular distance of 7″ in PA 338°. The o² Eri system is relatively nearby, just 16 light years away.

## α CVn Cor Caroli
### RA 12h 56.0m dec +38° 19′   Map 4

Appearing to the naked eye as a rather isolated 2nd-magnitude star between the tip of the Big Dipper's handle and the wedge of stars comprising the main body of Coma Berenices, α CVn is known as Cor Caroli (Charles's Heart, for King Charles II of England). Telescopic examination reveals an unevenly matched pair of white stars, mag. +2.5 and +5.6, separated by 19.4″ in PA 229°. These are easily split in my 80 mm refractor at ×40. The stars are physically connected, and lie at a distance of 92 light years.

## 61 Cyg
RA 20h 13.6m dec +38° 45′   Map 6

Fairly obscure as a naked-eye object, 61 Cyg is a 5th-magnitude star more or less at the right angle of a triangle it forms with Deneb and ε Cyg (the Swan's more easterly wing). The star is best known for being the first to have its parallax determined, in 1838 by Friedrich Bessel (1784–1846). A star's parallax is its apparent shift relative to more distant stars resulting from the Earth's orbital motion. From its parallax, 61 Cyg is found to lie only 11.2 light years away. Another clue to the star's proximity is its rapid motion against the stellar background (5″ per year to the NE). 61 Cyg is a good double star for small telescopes, with slightly orange components of magnitude +5.2 and +6.1, separated by 30″ at PA 146°.

## ζ Lyr
RA 18h 44.8m dec +37° 36′   Map 5

Marking the NW corner of the parallelogram of stars forming the outline of Lyra, ζ Lyr is an easily resolved pair with components of mag. +4.4 and +5.7 separated by 44″ in PA 150°. Even a 50 mm telescope, or steadily mounted 10 × 50 binoculars, should readily split them.

## γ Ari
RA 01h 53.5m dec +19° 18′   Map 7

Discovered to be a double star by Robert Hooke in 1664, this is one of the more closely matched pairs in the sky, with white components of mag. +4.6 and +4.7 separated by 7.8″ in PA 000° – a perfect N–S alignment.

## θ Ser
RA 18h 56.2m dec +04° 12′   Map 5

Tucked in to the south of Aquila's more westerly wing, θ Ser is a nicely matched pair of white stars of mag. +4.6 and +5.0, separated by 22.3″ in PA 104°. Observers using a medium power to sweep the rich Milky Way starfields in this region will find θ Ser an outstanding object. Despite the apparent close match, the two stars are unconnected: this is an optical double, with respective distances of 101 and 114 light years. To the naked eye, θ Ser appears as a single 3rd-magnitude star.

## ν Dra
RA 17h 32.2m dec +55° 10'   Map 1

An excellently balanced pair of mag. +4.9 stars, ν Dra marks the NW corner of Draco's head (an asterism sometimes called the Lozenge by American observers). Separated by 62″ (two Moon-widths) in PA 312°, this pairing of white stars is best appreciated in the low-power view of 7 × 50 or 10 × 50 binoculars or a finderscope.

## β Sco
RA 16h 05.4m dec −19° 48'   Maps 4, 5

Northernmost of the arc of 2nd-magnitude stars about 5° west of Antares, β Sco is an easy double for small telescopes, with components of mag. +2.6 and +4.9 separated by 13.6″ in PA 021° – as easy, say, as Mizar. These are young, luminous B-type stars, and the primary is a spectroscopic binary.

## β Cap
RA 20h 21.0m dec −14° 47'   Maps 5, 6

A wide optical double, ideal for binocular viewing, β Cap has a yellow-orange primary of mag. +3.1, with a pale green mag. +6.1 secondary 3.5' to its west. The respective distances of β¹ and β² Cap are 325 and 829 light years, so there is no physical connection. Each has been identified as a spectroscopic binary.

### Color Contrast Doubles

## β Cyg Albireo
RA 19h 30.7m dec +27° 58'   Map 5

The sky's finest wide color contrast double star is surely Albireo, marking the head of Cygnus at the end of the Swan's outstretched neck, a showpiece of the northern sky between June and September. Appearing to the naked eye as a single star of magnitude +3.1, Albireo splits in any small telescope into an orange mag. +3.3 primary and a green secondary of mag. +5.1, separated by 34.4″ in PA 054°. Even 10 × 50 binoculars will resolve the pair if steadily mounted. Some observers see

▶ Albireo is an easy, bright color contrast double, seen here in a photograph by Nick Hewitt.

the fainter component as bluish: color perception depends on the individual's eye and, in all probability, the equipment (the views in a refractor may be slightly redder than in a reflector, for example). Doubts have been raised as to whether Albireo is a genuine, gravitationally connected pair. The two stars do, however, share a common motion through space, and lie 410 light years away.

## o¹ Cyg
RA 20h 13.6m dec + 46° 44'   Maps 5, 6

Midway between Deneb (the Swan's tail) and δ Cyg lies an attractive triple arrangement of stars. o¹ Cyg is a K-type orange star of mag. +3.8. Located 4' away in PA 324° is 30 Cyg, which appears green to the eye and makes an attractive pairing in the low-power view. There is no physical connection between the two: o¹ Cyg lies 195 light years away, while 30 Cyg is at a distance of 270 light years. The attractiveness of the field is further enhanced by mag. +7.0 SAO 49338, a bluish star a couple of arcminutes to the east of o¹ Cyg.

## γ And Almach
RA 02h 03.9m dec +42° 20'   Map 7

▲ Closer together than those in Albireo, the components of the color contrast pair γ Andromedae make an attractive view in a medium-power telescope.

The southerly line of stars in Andromeda, trailing east from the Square of Pegasus, ends with a contender for the very best of the color contrast doubles. With colors as intense as those of Albireo, but components closer together (9.8″), γ And is a splendid telescopic object, comfortably resolved at magnifications of ×40 and upward. The primary is a K-type orange mag. +2.3 star, while the mag. +4.8 secondary (at PA 056°) is greenish. The secondary is itself double, with components of mag. +5.1 and +6.3 separated by only 0.5″, resolved only in very large telescopes, of 250 mm aperture and greater, and under the best conditions.

## ι Can
RA 08h 46.7m dec +28° 46'   Map 3

With a golden-yellow mag. +4.2 primary and a green mag. +6.6 secondary, separated by 30″ in PA 307°, ι Cnc resembles a fainter version of Albireo. Located about 8° due north from the cluster Praesepe (M44) at Cancer's center, this double is easily split in small telescopes and is an attractive object at a magnification of ×40.

## α *Her Rasalgethi*
### RA 17h 14.6m dec +14° 23'   Map 5

Known as a good color contrast double, α Her is a reasonably prominent naked-eye star 15° south of the Keystone in Hercules. The primary is an old red giant star which varies in brightness between mag. +3.1
and +3.9 as it pulsates. Telescopically, the primary's red color is obvious, and a green mag. +5.4 companion lies 4.6″ away in PA 107°; the
pair is easily split in an 80 mm aperture telescope. The companion star
is a spectroscopic binary, and the system lies about 400 light years away.

## γ *Del*
### RA 20h 46.7m dec +16° 07'   Map 6

Easternmost of the diamond of stars (sometimes called Job's Coffin)
that form the body of the Dolphin, γ Del, marking its nose, is reckoned by observers to be one of the sky's best color contrast doubles.
In his 1844 *Cycle of Celestial Objects*, Admiral Smyth described the two
components as "yellow and emerald green." I find the pair to be rather
muted in comparison to Albireo, though easily resolved (separation
9.6″) at ×40 in an 80 mm refractor. The stars have magnitudes of
+4.3 and +5.1.

## Multiple Systems

## θ¹ *Ori Trapezium*
### RA 05h 35.3m dec −05° 23'   Map 2

Located near the narrow end of the dark bay of the "Fish's Mouth,"
θ¹ Ori will be familiar to any observer who has enjoyed a high-magnification view of the Orion Nebula (p.99). Small telescopes show this as
a single star. My 80 mm refractor at ×40 will resolve the components,
though they can be quite difficult to separate clearly under some conditions – perhaps because of reduced contrast against the background

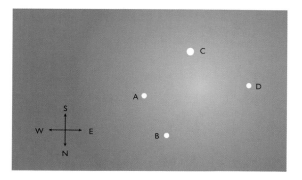

▶ *The components of the
Trapezium quadruple star
in the Orion Nebula are
lettered in order of RA.
This schematic view shows
them as they appear in
an inverted, astronomical
telescope view.*

nebulosity. In a 102 mm aperture instrument at ×100, resolution into the four components of the Trapezium is a lot clearer.

As its descriptive title suggests, the Trapezium consists of four stars, which are bound together gravitationally, and labeled A to D in order of increasing right ascension. The brightest is actually star C, at mag. +5.4 much more prominent than the other three. Stars A and D are each of mag. +6.7. Star B is usually at mag. +7.9 but is – like Algol – an eclipsing binary, undergoing fades to mag. +8.7 in a period of 6.47 days; it has a variable star designation as BM Ori. All four Trapezium stars appear white in the eyepiece.

These stars are the brightest of a cluster formed about 300,000 years ago in the heart of the Orion Nebula. Strong ultraviolet radiation from them illuminates the surrounding gas and dust. In time, radiation pressure from these vigorous, hot young stars will punch a hole through the nebula, and the cluster will then be revealed in its full glory, free from obscuration.

## β Mon
RA 06h 28.8m dec −07° 02'  Map 2

Located just north of the line from Orion's Belt to Sirius, β Mon is an attractive triple system consisting of three white A-type stars. The A–B pair have respective magnitudes of +4.6 and +5.4, and are separated by 7.3″ in PA 132°. A closer look shows the fainter star to have a closely matched, mag. +5.6 companion. The B–C pair has a separation of 2.8″ in PA 064°. This trio can be resolved comfortably in a 100 mm aperture telescope at ×100.

## ε Lyr Double Double
RA 18h 44.3m dec +39° 40'  Map 5

Together with the Trapezium, Lyra's Double Double is one of the best-known multiple stars in the sky, and a true showpiece object. Found just NE of the bright mag. 0 Vega, ε Lyr appears as a fairly tight naked-eye double (easily separated in binoculars) with 4th-magnitude components 3.5′ apart aligned more or less N–S. Each component is, in turn, found to be double in the telescope – a discovery made by William Herschel in August 1779.

The northern pair ($\epsilon^1$/A–B) is mag. +5.4 and +6.5, with a separation of 2.6″ at PA 173°, orbiting in a period of 600 years. The southern pair ($\epsilon^2$/C–D) is slightly brighter and tighter, at mag. +5.1 and +5.3, and separated by 2.3″ in PA 094° with a 1200-year orbital period.

All four stars in this system are white and gravitationally bound together. They lie about 160 light years away, the two pairs being 0.16 light year from each other.

Theoretically, a 75 mm aperture telescope should comfortably split the two pairs. I find, however, that the Double Double, neatly contained in the field at ×150, is better separated in a 100 mm or 150 mm aperture instrument.

## ν Sco
### RA 16h 12.0m dec −19° 28′  Maps 4, 5

Lying a couple of degrees to the east of β Sco (p.139), ν is a multiple system. A small telescope shows it as a pair with components of mag. +4.2 and +6.1, separated by 41.1″ at PA 337°. The brighter component is seen in large telescopes to itself be double, with stars of mag. +4.6 and +5.2 separated by 1.2″.

## Some Challenging Pairs

## ζ Ori Alnitak
### RA 05h 40.8m dec −01° 57′  Map 2

Southeasternmost of the three stars in Orion's Belt, ζ Ori is a good, testing double star for telescopes in the 70 to 100 mm aperture bracket. The mag. +4.0 secondary lies 2.4″ in PA 162° from the mag. +1.9 primary. In my 102 mm refractor, I found that I needed a magnification of ×200 on a night of average seeing to split this pair of white stars.

## α Gem Castor
### RA 07h 34.6m dec +31° 53′  Maps 2, 3

Castor is the more northerly of Gemini's "twin" stars, and a celebrated multiple system. Its nature as a double star was discovered in the seventeenth century, while William Herschel was the first to detect movement in the relative position of the A and B components, in 1804. The A–B pair have an orbital period of 470 years, and were last at periastron in 1965. At a separation close to 3″ in PA 171° (roughly E–W), and with respective magnitudes of +1.9 and +2.9, Castor A and B can be split in good conditions with an 80 mm refractor. Both stars appear white.

A fairly distant companion, mag. +9.1 Castor C, lies 72″ away in PA 164°. Castor A, B and C have each been found to be spectroscopic binaries; A and B are each pairs of A-type stars, while Castor C is a pair of red dwarfs. Castor is thus a sextuple system, and lies 48 light years from us.

## ξ UMa
### RA 11h 18.2m dec +31° 32′  Map 3

One of the fainter stars making up the hind legs of the Great Bear, about 10° north of Leo's tail, ξ UMa is historically noteworthy in being

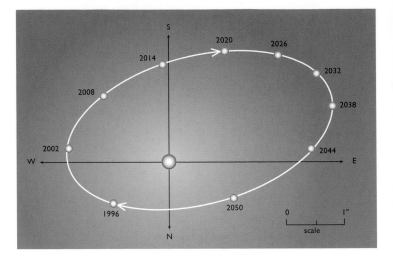

▲ ξ Uma shows rapid changes in both separation and position angle (PA), as illustrated above, thanks to its relatively short-period orbit.

the first binary star to have its orbit (59.74 years) calculated. Its binary nature was established by William Herschel in 1789. ξ UMa consists of a pair of Sun-like yellowish G-type stars, separated by just under 2″, at a distance of 26 light years.

## γ Leo Algieba
### RA 10h 20.0m dec +19° 51′  Map 3

The third star working northward in the Sickle of Leo, Algieba is a good, reasonably close pair (separation 4.4″, PA 122°) for small telescopes, with white components of mag. +2.2 and +3.5. It lies at a distance of 75 light years. In an 80 mm telescope at ×100, I find Algieba elongated but not split: a 102 mm telescope, however, clearly shows dark sky between the rather unequal components at the same magnification.

## ε Boo Izar
### RA 14h 45.0m dec +27° 04′  Map 4

The second-brightest star in Boötes is a good, challenging double for small telescopes, with a yellow mag. +2.7 primary and subtly greenish mag. +5.1 companion separated by 2.8″ in PA 339°. This is another pairing where the relatively bright primary rather overwhelms the secondary in small instruments, especially on nights of unsteady seeing. I find it quite testing even at ×200 in a 102 mm refractor.

## α *Sco Antares*
RA 16h 29.4m dec −26° 26'   Maps 4, 5

The famous red giant heart of the Scorpion has a catalog magnitude of +1.0, and is found on careful telescopic inspection to have a greenish mag. +6.5 companion, 3.0″ away in PA 273°. The large difference in brightness between the components can make this a tricky double to resolve, especially when seen low down in unsteady seeing conditions.

## ζ *Aqr*
RA 22h 28.8m dec −00° 01'   Map 6

At the fork of the Y-shaped Water Jar asterism in Aquarius, ζ Aqr is a well-matched pair of white stars, of mag. +4.4 and +4.5, separated by 2.1″ in PA 200° – a good test for 80 to 100 mm aperture telescopes. The pair lie at a distance of 98 light years, and have an orbital period of 760 years.

## α *Psc*
RA 02h 02.0m dec +02° 46'   Map 7

With a separation of 1.8″ in PA 050°, and components of mag. +4.3 and +5.2, this is another testing target for instruments in the 100 to 150 mm aperture class.

## ε *Ari*
RA 02h 59.2m dec +21° 20'   Map 7

Even more perfectly matched than γ Ari, ε presents a pair of white mag. +4.6 stars separated by 1.4″ in PA 208°, a very testing target for 100 to 150 mm aperture telescopes.

## α *Cru Acrux*
RA 12h 26.6m dec −63° 06'   Map 8

The bright star at the southern end of the Southern Cross is a marvellous double for 75 mm and larger telescopes, with blue-white B-type components of mag. +1.3 and +1.7 separated by 4.4″ in PA 115°. The system lies 321 light years away.

# 8 · PLANETARY NEBULAE

After they have formed, stars of similar mass to our Sun will shine steadily for billions of years as a result of the process of nuclear fusion in their core. For most of their lives, the principal fusion reaction in such stars combines four atoms of hydrogen to produce a single helium atom, with the release of energy and photons of light. Eventually the core hydrogen becomes depleted, and fusion reactions begin which convert helium to carbon. During this phase the star will swell to become a red giant, losing material from its distended outer atmosphere into interstellar space through a strong stellar wind. Later still, the outer layers of the star are expelled to form a planetary nebula, with the hot core exposed as a dwarf star at the center of the nebula. The hot central star (surface temperature 100,000 K; for comparison the Sun's photosphere has a temperature of 6000 K) emits strongly in the ultraviolet part of the spectrum, ionizing gas in the tenuous surrounding envelope.

Planetary nebulae are short-lived on the cosmological timescale, lasting perhaps a few tens of thousands of years before diffusing into the interstellar medium. For this reason they are not especially common: about 1500 are known in our Galaxy. Bright examples have been found in nearby galaxies; more than a hundred have been detected in the Magellanic Clouds, while planetary nebulae are also known in the Andromeda Galaxy (M31) and in two of its outlying satellites, NGC 147 and NGC 185. Planetary nebulae enrich the interstellar medium with heavier elements synthesized inside their progenitor stars.

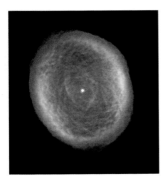

▲ IC 418, the Spirograph Nebula in Lepus, shows exquisite detail sculpted by magnetic fields, as seen in this image from the Hubble Space Telescope.

Long-exposure images from large professional telescopes show planetary nebulae to have greatly extended envelopes, up to a couple of light years in size and with intricate structures. Shock waves can be seen where material more recently ejected at high velocities (15–20 km/s) piles into the slower-moving gas lost during the star's earlier red giant phase. Hubble Space Telescope images of the Helix Nebula (NGC 7293) in Aquarius have revealed comet-like radial features presumed to result from the presence around the central star of sizable solid bodies which are being eroded by the material ejected from the outer layers. The Eskimo Nebula (NGC 2392) in Gemini, and the Spirograph Nebula (IC 418) in Lepus show delicate structure, sculpted by their central stars' magnetic fields, in HST images.

The term "planetary nebula" was first introduced by William Herschel in 1785, four years after his discovery of Uranus. Herschel likened the appearance of these objects to fainter versions of the dim, sixth-magnitude planet: they looked, in his telescopes, like extended faint disks.

Much of the light in planetary nebulae comes from oxygen atoms which have been stripped of two electrons. When such doubly ionized atoms capture free electrons from the surrounding medium, the quantum energy budget is balanced by emission of light at (principally) 500.7 nm and at 485.9 nm. This O III emission is toward the green end of the spectrum, and accounts for the greenish tints sometimes reported by visual observers.

The nature of planetary nebulae as objects shining by emission (rather than by reflection of starlight) was established through spectroscopic observations by William Huggins in 1864. At the time of this discovery, O III emission was though to be produced by a previously unknown chemical element – dubbed "nebulium" – unique to the nebulae. It was later established that this emission was from oxygen in a very rarefied state: the O III emission is described as "forbidden," since it does not occur under terrestrial laboratory conditions.

The predominance of O III emission in the light of planetary nebulae can be used to advantage by observers. Narrow-passband filters, which allow through only the 500.7 and 485.9 nm wavelengths of O III, can enhance the contrast between the planetary nebula and the sky. Very little starlight is emitted at these wavelengths, so an O III filter will markedly dim the stars in the field, while having little effect on the visibility of the nebula itself. Owners of large telescopes can use an O III filter to identify very distant, almost point-like planetary nebulae, which will remain visible through the filter while stars become dim.

An O III filter can either be installed in the conventional screw-thread housing at the field end of the eyepiece, or hand-held between the eye lens and the eye. The second option allows the observer to alternate rapidly between filtered and unfiltered views, which I find works well for nebulae such as the Ghost of Jupiter (NGC 3242). An O III filter is not essential for planetary nebula observing – many of these objects are well seen without such aid – but it can be interesting to compare the appearance in integrated light with the filtered view.

UHC filters, which pass the hydrogen-β as well as the O III emission lines, are also useful for observing planetary nebulae.

There is no hard-and-fast classification system for the appearance of planetary nebulae. The uniqueness of each of these objects is perhaps reflected in the abundance of popular names for them – Hourglass, Egg, Ant, Eskimo, and so on. In some planetary nebulae the central star is more prominent than in others: in many cases, such as the famous Ring Nebula (M57), it may be invisible in small telescopes. Planetaries differ considerably in outline. Some, like M57, are well-defined rings, while others may be more disk-like – the Owl Nebula (M97), for example. The nebulosity may be in two lobes to either side of the central star, as in the Dumbbell Nebula (M27).

These differing shapes are indicative of the processes by which planetary nebulae are formed. In some cases, the ejection of gas is in strong polar outflows, governed by the star's magnetic field, which produces symmetrical lobes of material. In others the ejection is in all directions, producing "bubbles" which appear more uniform or disk-like.

Since planetary nebulae represent a short-lived phase of stellar evolution, astronomers have limited opportunities to study the formation process in action. The variable star FG Sagittae, surrounded by material it ejected 6000–10,000 years ago, is believed to be an example of a nascent planetary nebula; HM Sagittae may be another.

Nearby planetary nebulae, naturally, tend to have a larger apparent size. This can be a disadvantage from the observational point of view, as such objects have low surface brightness and poor contrast with the sky. For instance, the Helix Nebula (NGC 7293) covers a large area of sky but is difficult to see. Remote planetaries, on the other hand, appear almost star-like, revealing their true nature only when examined with an O III filter in place.

In general, planetary nebulae are best seen telescopically. Only the Dumbbell Nebula shows reasonably well in typical amateur astronomers' binoculars: the others are too small to be distinguishable from stars at low magnifications.

Hunting down these objects telescopically will often require the use of averted vision, bringing in another interesting effect. Some planetary nebulae are described as "blinking." These tend to be objects where the central star is relatively bright in comparison with the nebula. Sometimes, with direct vision, only the star shows up well, but when averted vision is used the nebula blinks back into view. This happens in particular with NGC 6826, which has consequently become known as the Blinking Planetary.

Observing planetary nebulae, which are faint and of low surface brightness, demands very dark, transparent conditions. Under the best conditions, when the seeing is steady, it may be possible to pick out some structure in brighter, reasonably large planetaries such as M57 or M27 in amateur telescopes. As with other objects, it is always worth making a field sketch by way of record, and the visibility or otherwise of the central star can be noted. Particularly when observing at higher magnifications, it is also worth noting the steadiness of the seeing conditions on the Antoniadi scale (Table 4), as this can have an influence on the appearance of some planetary nebulae.

| TABLE 4: THE ANTONIADI SEEING SCALE | |
| --- | --- |
| I | Perfect seeing without a quiver |
| II | Slight undulations; moments of calm lasting several seconds |
| III | Moderate seeing, with larger air tremors |
| IV | Poor seeing, with constant troublesome undulations |
| V | Very bad seeing; even a rough sketch impossible |

## Selected Planetary Nebulae

### M27 NGC 6583 Dumbbell Nebula
Vulpecula   RA 19h 59.6m dec +22° 43′   mag. +7.3   Maps 5, 6

The brightest and best of the planetary nebulae for small amateur telescopes has to be M27, the Dumbbell Nebula in Vulpecula. This object is easily found in binoculars, a short 3° hop due north from γ Sge, the star at the arrow's tip. With angular dimensions of 8′ × 5′ (the longer axis is aligned NE–SW), the Dumbbell shows as a small

▲ The brightest of the planetary nebulae, M27 is a good target even for binocular observers. It is visible to left of center in this photograph by Nick Hewitt.

hazy patch in binoculars. The brightest nearby field star, at mag. +5.5, is 14 Vul, 20' to the north of M27.

Telescopes as small as 80 to 100 mm aperture show some hints of the structure familiar from long-exposure images. M27 has two lobes of nebulosity connected by a narrow waist – an appearance often described as resembling an apple core. Observers using 150 mm aperture telescopes and larger may glimpse some filamentary structure within the nebulosity, which appears grayish to the eye. An O III filter will enhance this fine detail and make the nebula appear slightly larger and rounder.

M27 was discovered by Messier in July 1764. It lies 1000 light years away and has a hard-to-see 14th-magnitude central star with a surface temperature of 85,000 K.

## M57 NGC 6720 Ring Nebula
Lyra   RA 18h 53.6m dec +33° 02'   mag. +8.8   Map 5

M57 is one of the easiest planetary nebulae to locate, lying roughly midway between the 3rd-magnitude β and γ Lyr at the south of the parallelogram of stars making up the lyre's body. Familiar from countless images reproduced in books and magazines, this is one of the few planetary nebulae whose telescopic appearance lives up to the photographs – at least in reasonable-sized instruments with a bit of magnification – a 150 mm reflector at ×200, say.

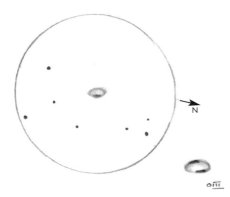

▲ The Ring Nebula (M57) in Lyra as seen in the author's 102 mm refractor at ×100. Use of an O III filter reveals slightly more structure in the main, bright ring of nebulosity, but the central star is beyond reach for small amateur telescopes.

With an angular diameter of 70″, M57 is just too small to be distinguishable from the stellar background in binoculars: a small telescope magnifying about 20 times is perhaps the minimum requirement to show the Ring Nebula's non-stellar oval outline. Low powers in small telescopes show this fairly high surface brightness object as an even disk. A magnification of ×40 gives more of an impression of a brighter outer ring with a fainter interior, while at ×80 the Ring Nebula certainly lives up to its popular name in my 80 mm refractor.

In larger telescopes, this is a breathtaking object – a gray-white hoop of light which may appear to shimmer like a celestial smoke-ring on nights of unsteady seeing. Careful examination, aided by an O III filter, can bring out subtle brightness variations within the nebulosity.

The Ring Nebula's central star is of magnitude +15.3, and very difficult to see even in a large telescope. (I once had a chance to look for it in a 510 mm reflector, and could not see the star.) The nebula has an estimated age of 6000 years. M57 was discovered in January 1779 by the French astronomer Antoine Darquier (1718–1802).

## NGC 2392 Eskimo Nebula
Gemini   RA 07h 29.2m dec +20° 55′   mag. +9.2   Map 2

This is another planetary often portrayed in magazines and books – the January 2000 Hubble Space Telescope view (p.9) is especially well known. NGC 2392 has a high surface brightness, making it one of the easiest planetary nebulae for amateur observers. Popularly known as the Eskimo Nebula (or, particularly in the United States, the Clown Face Nebula), NGC 2392 can be found by quite a simple star-hop about 3° WSW from the mag. +3.5 star δ Gem. En route, look for the mag. +5.2 star 63 Gem, at the southern tip of a flat triangle it forms with a couple of 6th-magnitude stars, then move half a degree to the SSE, where a low-power view shows what appears to be an evenly matched pair of 9th-magnitude stars, separated by a couple of arcminutes. The more southerly is, in fact, the Eskimo, and use of higher magnifications will make this object look increasingly "fuzzy:" unlike the star to its north, NGC 2392 cannot be brought to a sharp focus.

Small telescopes show the bright, 15″-diameter central part of NGC 2392; the outer, dimmer surrounding shell requires larger instruments. In my 80 mm refractor, I find the Eskimo's core to be markedly bluish at ×40 and ×80. A low-power view through an O III filter dims down the field stars, making NGC 2392 stand out from the neighboring 9th-magnitude star.

In larger telescopes (100 mm aperture and greater) the outer halo – the hood of the Eskimo's parka in long-exposure images – becomes visible, particularly with an O III filter in place. NGC 2392 certainly appears markedly larger with an O III filter than with-

▲ NGC 2392, the Eskimo Nebula, appears as the more southerly of a pair of 9th-magnitude "stars" in the low-power view. Closer examination reveals the planetary as diffuse, as seen in this sketch from the author's 80 mm f/5 refractor at ×40.

out at ×200 in my 102 mm refractor. NGC 2392 was discovered by William Herschel in 1787. It has a mag. +10.5 central star.

## M97 NGC 3587 Owl Nebula
Ursa Major   RA 11h 14.8m dec +55° 01'   mag. +9.9   Map 1

Sometimes described as the most difficult of all the Messier objects (though I would rate M76, described below, as harder to see), M97 has a very low surface brightness, making its 3' to 4' circular outline hard to detect except under very dark, transparent conditions. For observers in the northern hemisphere, the best time to look for this object is during the spring months of March and April, as Ursa Major wheels high into the eastern sky. M97 is found 2.5° SE of Merak (β UMa, the southerly "Pointer" in the Big Dipper). An easy star-hop from Merak brings the observer to a loose L of 7th-magnitude stars; M97 is about 10' ENE of the star at the tip of the L's longer arm. Averted vision makes M97 stand out reasonably well in an 80 mm telescope.

M97's popular name, the Owl Nebula, comes from observations made by Lord Rosse with his 72-inch telescope in 1848, in which he drew attention to a couple of darker patches – "eyes" – on the brighter background of the nebula. These are visible in large amateur instruments, but are beyond the reach of telescopes in the 80 to 150 mm aperture class. Discovered by Méchain in 1781, M97 has a 16th-magnitude central star, and lies 2600 light years away.

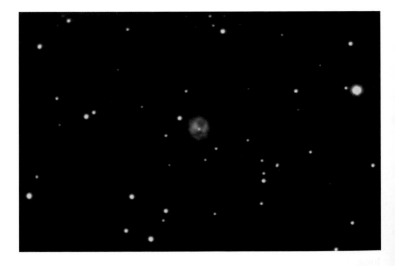

▲ The Owl Nebula (M97) has a circular profile with two dark patches. These "eyes" are visible in long-exposure images, such as this taken by Geoffrey Johnstone.

While in the Owl's neighborhood, it is worth seeking out another Messier object, the almost edge-on 10th-magnitude galaxy M108 (NGC 3556) at RA 11h 11.5m, dec +55° 40′, just under 50′ away in the same low-power field as the Owl.

## M76 NGC 650/651 Little Dumbbell
Perseus   RA 01h 42.4m dec +51° 34′   mag. +10.1   Maps 1, 7

Widely considered to be one of the more difficult Messier objects for observers using small telescopes, M76 lies just inside the NW border of Perseus. The best guide to hunting it down is the mag. +4.1 star φ Per, roughly midway between γ And and α Cas: M76 lies 40′ to the NNE of φ Per. Although quite extended (3′ × 1′ in angular size), M76 doesn't really show well at low powers. I find that in my 80 mm it is visible at ×40 but not at ×20; perhaps the slightly darker field at the higher magnification helps improve the contrast between this low surface brightness object and the sky? A 7th-magnitude star lies a couple of arcminutes due east of the nebula. In a small telescope, I see M76 best by using averted vision: indeed, switching between averted and direct vision makes this planetary "blink" when I look at it. The nebula appears oval, and with the long axis lying NW–SE.

M76 was discovered by Méchain in September 1780. It has the popular nickname of the Little Dumbbell, though it is a good deal more difficult to see than the Dumbbell itself, M27. Like M27, it shows a couple of connected lobes of nebulosity; each has been given its own designation in the NGC. At mag. +15.9, the central star is beyond the reach of small amateur telescopes. Also known as the Butterfly Nebula, M76 lies at a distance of 3400 light years.

## NGC 6543 Cat's Eye Nebula
Draco   RA 17h 58.6m dec +66° 38′   mag. +8.1   Map 1   Finder chart p.154

Another planetary nebula made familiar by a (1995) Hubble Space Telescope image, NGC 6543 is one of the most strongly colored of these objects, frequently described as blue or blue-green by visual observers. Popularly known as the Cat's Eye Nebula, NGC 6543 emits strongly at O III wavelengths, and it was by observing this object that Huggins first determined the emission nature of planetary nebulae.

The Cat's Eye is roughly midway between ζ and δ Dra in a rather sparse part of the northern sky. The pair of 6th-magnitude stars 37 and 38 Dra are a useful guide. A 2° hop SSW from this pair takes the observer to a stretched quadrilateral of 8th-magnitude stars about 40′ long, lying NNE–SSW. A further 20′ in the same direction is a 7th-magnitude star. A line extending a further 20′ to the SSW takes the observer to NGC 6543. In a low-power view, I find that the Cat's Eye

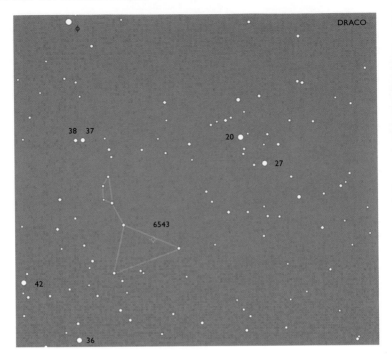

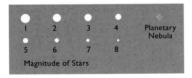

▲ Finder chart for NGC 6543, the Cat's Eye Nebula. It shows stars to limiting magnitude +8.5 in a field with diameter 8°.

looks like a decidedly blue 8th-magnitude star. The difference in visibility in an O III filter is even more dramatic than for the Eskimo Nebula – NGC 6543 shines strongly in this light.

While NGC 6543 has an overall circular angular extent of about 6′, only the central region, spanning 25″, is readily seen by visual observers. The nebula's core has a relatively high surface brightness. Immersed in nebulosity, NGC 6543's mag. +10.9 central star is hard to see, except in larger telescopes, and in these it is lost when an O III filter is used to bring up the contrast of the nebula. The object lies 3000 light years away.

## NGC 3242 Ghost of Jupiter
Hydra   RA 10h 24.8m dec −18° 38′   mag. +7.8   Map 3

Among the brighter planetary nebulae, at mag. +7.8, NGC 3242 is often said by visual observers to have a pronounced blue or blue-green color. NGC 3242 is easy to locate, nearly due south of the 4th-magnitude star μ Hya, and 15° ESE from Alphard (α Hya). Just over a degree south of μ Hya, a line of three 7th to 8th-magnitude stars runs SW for about 40′; closer examination shows the star at the NE of the line to be a pair, while that at the SW is in fact NGC 3242.

The planetary can be seen as a point of light in 10 × 50 binoculars, and despite its relatively low altitude in UK skies (from my location in southern England it culminates 20° up), I find this a very easy object in my small 80 mm refractor. Strongly blue and star-like at ×20, it swells into an extended disk with a sharp core at ×80, showing perhaps a slight offset of its brightest region toward PA 170°. An O III filter really brings up the contrast between this object and the sky.

NGC 3242 gets its popular name of the Ghost of Jupiter from its dim, washed-out disk with a diameter of 40″, similar to that of the giant planet. Larger telescopes show the core to be surrounded by a darker region, enclosed by the brighter outskirts of the nebula. NGC 3242 appears somewhat elongated E–W in high-power views. The Ghost of Jupiter has a mag. +12.1 central star and lies 2000 light years away. It was discovered by William Herschel in 1785.

## NGC 6572

Ophiuchus   RA 18h 12.1m dec +06° 51′   mag. +8.1   Map 5

NGC 6572 is a small, bright, greenish planetary nebula showing a 15″ disk in 100 to 150 mm aperture telescopes. Lying in a rather barren field, it can be found 2.5° SSE of 4th-magnitude 72 Oph, or 5° NNE from the triangle of 67, 68 and 70 Oph. The central star of NGC 6572, at mag. +12.9, is invisible in small amateur telescopes.

## NGC 6826 Blinking Planetary

Cygnus   RA 19h 44.8m dec +50° 31′   mag. +8.8   Maps 5, 6   Finder chart p.156

NGC 6826, near θ Cyg on the Swan's westerly wing, is known as the Blinking Planetary. As a result of the relative brightness of the (mag. +10.6) central star and nebula (overall mag. +8.8), this planetary appears and disappears when averted and direct vision are used alternately; in the

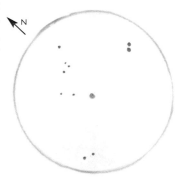

▶ The author's sketch of the Blinking Planetary (NGC 6826), as seen in an 80 mm f/5 refractor at ×40. The bright double star to the west of NGC 6826 is 16 Cygni, which is a good guide for finding the object.

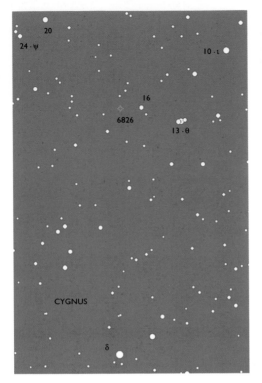

◄ *Finder chart for the Blinking Planetary (NGC 6826) on the western wing of Cygnus. The chart shows stars to limiting magnitude +8.5 in a 5.25°-wide field.*

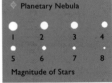

direct view, the star overwhelms the nebulosity. NGC 6826 is quite bright, and lies in range of 80 to 100 mm aperture telescopes.

## NGC 7026

Cygnus   RA 21h 06.3m dec +47° 51'   mag. +10.9   Maps 1, 6

A tricky object in low-power views, with several relatively bright stars in its field, NGC 7026 is located 11' north of the mag. +4.6 star 63 Cyg and 5° ENE of Deneb. The planetary nebula lies adjacent to a 7th-magnitude star. A 150 mm aperture telescope will show a bluish-green disk 15" across, but no hint of the mag. +14.8 central star.

## NGC 7662 Blue Snowball

Andromeda   RA 23h 25.9m dec +42° 33'   mag. +8.3   Maps 6, 7

Located in the northern reaches of Andromeda, 3° west of the 4th-magnitude star ι And, NGC 7662 has an apparent diameter of 12", making it one of the most compact planetary nebulae within the grasp of smaller amateur telescopes. Fairly bright at mag. +8.3, this object's popular name of the Blue Snowball is indicative of its marked color for visual observers. Use of an O III filter can help to identify it in a low-power field. Higher-magnification views show a bright core surrounded by a fainter, more diffuse "halo."

## NGC 7009 Saturn Nebula
Aquarius RA 21h 04.2m dec −11° 22′ mag. +8.3 Map 6

NGC 7009 is fairly close to the pair of Messier objects M72 and M73. Indeed, the latter can be used as a guide for finding the planetary, 5° ESE of ε Aqr: NGC 7009 is a degree north of M73, and 5′ of RA to its east, so placing M73 (an asterism described on p.131) to the south of a 1.5° field, and leaving the telescope stationary for 5 minutes to allow Earth's rotation to "move" the stars, will bring NGC 7009 into the field of view. Observers in the British Isles may find NGC 7009 a little low in the sky, but at mag. +8.3, it should show reasonably well in 100 to 150 mm aperture instruments on a good night.

NGC 7009 has quite a flattened, oval outline, with dimensions of 44″ × 26″, the long axis lying E–W. Lord Rosse named it the Saturn Nebula in 1850, from observations which showed extensions – "ansae" – to either side, rather like the extremities of Saturn's rings. NGC 7009 has similar angular dimensions to the planet as seen with its rings open in their presentation toward Earth. The ansae are beyond the reach of small telescopes, as is the 12th-magnitude central star.

## NGC 7293 Helix Nebula
Aquarius RA 22h 29.6m dec −20° 48′ mag. +7.3 Map 6

At a distance of 300 light years, the Helix is the closest of the planetary nebulae, and with an integrated magnitude of +7.3 might be expected to be an easy object. Its light is, however, spread over an area

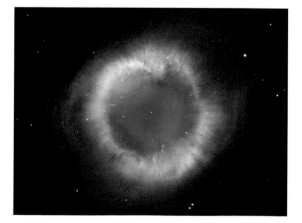

▶ Visually rather dim and difficult, the large, low-contrast Helix Nebula is a familiar photographic subject, captured here by the Hubble Space Telescope.

13′ across – getting on for half the Moon's apparent width – giving it an extremely low surface brightness. This makes it a really elusive object, and its southerly declination and low elevation in North American and UK skies also render it hard to see. Observers in the southern hemisphere at least get the advantage of having the Helix Nebula – another planetary familiar from long-exposure images – well up in the sky. A low-power, wide-field view, aided by an O III or UHC filter, perhaps gives the observer the best chance of seeing NGC 7293; it reportedly does not take high magnification well, but a filtered view in an aperture as small as 80–100 mm can be remarkably good. Many experienced deep sky observers have never seen the Helix! It was discovered by the noted German asteroid-searcher Karl Ludwig Harding (1765–1834).

## NGC 2438
Puppis   RA 07h 41.8 dec −14° 44′   mag. +11   Maps 2, 3

Located in the foreground of the 5400 light years distant rich open cluster M46 (p.124), NGC 2438 is 3300 light years away. Its position, 7′ north of the center of M46, makes it easy to find despite its faintness. A 100 mm aperture telescope will show it reasonably well. In a 150 mm at ×200, NGC 2438 stands out as a small, even, circular haze against the background stars (the planetary's mag. +17.5 central star is well beyond reach). NGC 2438 has an apparent diameter of about an arcminute.

## Some Challenging Planetaries

## NGC 2440
Puppis   RA 07h 41.9m dec −18° 13′   mag. +9.4   Maps 2, 3

Puppis' other reasonably noteworthy planetary nebula is NGC 2440, located in a rich Milky Way starfield 3.5° due south of M46. A faint, difficult oval object with maximum diameter of 32″ (the long axis running NW–SE), it requires at least a 100 mm aperture telescope and very transparent conditions.

## NGC 1535
Eridanus   RA 04h 14.2m dec −12° 44′   mag. +9.6   Maps 2, 7

Lying 5° due south of 4th-magnitude pair o¹ and o² Eri, and 4° ENE of 3rd-magnitude γ Eri, NGC 1535 is a small, circular planetary nebula in a rather empty part of the sky. Visible in a 100 mm aperture instrument, this object has a relatively bright, mag. +11.6, central star, surrounded by a strong halo of nebulosity. In telescopes of 150 mm aperture and greater, NGC 1535 shows a pronounced blue color. An

O III filter enhances its appearance, and used with a very large amateur telescope will reveal an outer ring of nebulosity.

**NGC 6891**

Delphinus   RA 20h 15.2m dec +12° 42'   mag. +10.5   Maps 5, 6

About 4.5° WNW of ε Del (the Dolphin's tail), NGC 6891 is a small, round disk 10″ in diameter. The central star is visible in a 150 mm aperture telescope.

**IC 3568**

Camelopardalis   RA 12h 32.9m dec +82° 33'   mag. +10.6   Map 1

In a rather empty part of the far northern sky east of Ursa Minor's tail (the nearest reasonably prominent field star at mag. +5.9 is 32 Cam, a degree to the NNW), IC 3568 is a rather small planetary nebula with an apparent diameter of 6″. In small telescopes it appears almost starlike and may require use of an O III filter for positive identification. In larger telescopes, IC 3568 appears bluish.

# 9 · SUPERNOVA REMNANTS

Whereas stars like our Sun end their days quite sedately as slowly fading white dwarfs, following a brief planetary nebula phase, those which are more massive make dramatic exits. Stars with a mass greater than the Chandrasekhar limit (1.4 times the Sun's mass) are destined to end their life in a supernova explosion, during which they will briefly shine with as much light as all the other stars in their host galaxy combined. The supernova explosion throws material into the interstellar medium and may destroy the star, leaving behind an expanding patch of nebulosity – a supernova remnant.

Supernovae occur by two principal mechanisms. In one, a hydrogen-depleted dwarf star in a close binary system accumulates material from a giant companion, until the Chandrasekhar limit is exceeded, leading to its collapse and runaway nuclear reactions which eventually cause a catastrophic explosion. Supernovae of this kind are described as Type I.

Since all Type I supernovae occur by the same mechanism, they are essentially identical events, all having the same maximum light output. This makes them useful as "standard candles" for estimating the distances to remote galaxies. Astronomers know the theoretical peak luminosity for Type I supernovae, so by accurately measuring the apparent magnitude they can determine the distance, based on how much the supernova's light has been attenuated.

Type II supernovae mark the demise of massive single stars, and are commonly found in the spiral arms of galaxies among star-forming H II regions with their attendant OB associations (p.111). Massive, highly luminous stars like Rigel in Orion have short hydrogen-burning stages, measured in millions of years, as opposed to billions of years for less profligate, Sun-like stars. Once the bulk of the hydrogen has been consumed, helium fusion takes over, with the star swelling to become a red supergiant like Betelgeuse, while its core contracts and becomes hotter.

Progressively heavier elements undergo fusion to maintain the star's energy output, but with steadily diminishing returns. As the end nears, the core of a supergiant star has a layered, onion-like structure, with iron at the center, surrounded by successive shells of silicon, neon, oxygen, carbon, helium and hydrogen. Fusion of silicon to iron occurs in the star's final moments, and once the mass of iron reaches the Chandrasekhar limit the core collapses abruptly, no longer able to support itself (fusion of iron to heavier elements requires more energy than it produces). The core's collapse sends shock waves through the outer layers of the star, triggering fusion of silicon and sulfur to produce radioactive cobalt and nickel, and is also accompanied by the release of a pulse of neutrinos. Meanwhile, the star's outer layers fall inward under gravity, because they are no longer supported by radiation pressure from the core. When it hits the core, this material rebounds, tearing the star apart in a supernova explosion.

Depending on the mass of the progenitor star, core collapse in a Type II supernova will leave behind either a black hole (if the star's original mass was greater than five times the Sun's) or a rapidly spinning, extremely dense and compact neutron star (if it was less than 5 solar masses). Intense magnetic fields around a neutron star can produce beams of radiation which can be detected, if favorably oriented in our line of sight, in radio telescopes; neutron stars thus detected are called pulsars.

From observations of their frequency in other galaxies, astronomers estimate that supernovae should occur at intervals of about 30 years in the Milky Way. However, the last such event witnessed in our Galaxy was in 1604; this was known as Kepler's Star, in Ophiuchus, named for its discoverer, Johannes Kepler (1571–1630). It seems

▲ *Supernova 1987A in the Large Magellanic Cloud was the brightest, and closest, such event since the invention of* *the telescope. The presupernova field is at right. This was a Type II event involving the demise of a massive single star.*

likely that any Galactic supernovae since then have been obscured from view by intervening gas and dust clouds in the Milky Way's spiral arms. Records of supernovae in antiquity have been found among Chinese and Korean annals. Seven definite supernovae in our Galaxy in the last 2000 years can be identified from records; the four hundred years since the last recorded Galactic supernova is not actually that unusual.

Supernovae in our Galactic neighborhood are brilliant events indeed. The 1572 supernova in Cassiopeia was discovered by the Danish astronomer Tycho Brahe (1546–1601) on November 6, and became known as Tycho's Star. It rivaled Venus at its brightest, and could be seen in daylight; it remained visible to the naked eye until March 1574. Chinese and Korean records suggest that the 1006 event in Lupus was comparable in brightness to the Moon at first quarter! Kepler's Star reached a magnitude of $-3$.

The brightest supernova in modern times was seen to erupt on February 24, 1987, in the Large Magellanic Cloud, close to the Tarantula Nebula (p.103). Designated SN 1987A, it reached a peak magnitude of $+3$ some three months later. Its blue giant progenitor star was identified on archive photographs of the LMC, and the gradual development of its expanding remnant is still being followed by professional astronomers using large telescopes. The part of the LMC in which SN 1987A exploded is a region of ongoing starbirth, and the remnants of several past supernovae can be seen nearby in long-exposure images.

Supernovae in other galaxies are detected by amateur astronomers equipped with relatively advanced CCD cameras and large telescopes who "patrol" dozens of galaxies in search of them each possible clear night. Early alerts of new stellar "interlopers" not present on archive images can allow professional astronomers to obtain valuable spectroscopic and other data.

From the point of view of the amateur deep sky observer, supernova remnants are interesting, if exceedingly scarce targets. Only the Crab Nebula and the Veil Nebula are reasonably accessible for observers using small telescopes in the northern hemisphere, while those in the southern hemisphere can see the brightest part of the Vela Supernova Remnant (SNR). The Crab is an easy enough object, but the main challenge with both the Veil and the Vela SNR lies in seeing them in the first place. It can be interesting to compare drawings made at the eyepiece with photographs and CCD images. An O III or UHC filter can help to bring the best out of these objects. For the tenuous wisps of the Veil in particular, the clearest and darkest conditions are essential.

## The Crab Nebula

### M1 NGC 1952 Crab Nebula
Taurus   RA 05h 34.5m dec +22° 01′   mag. +8.4   Map 2

The brightest of the supernova remnants, M1 is easy to locate, lying a degree NW of the 3rd-magnitude star ζ Tau (the Bull's more southerly "horn"). On a reasonably dark, clear night it is visible as a small, rather undistinguished hazy patch in 10 × 50 binoculars. A small telescope with a little more magnification will greatly improve the view, and give a better impression of shape and size. At ×20 in a wide-field 80 mm refractor, M1 appears as a uniformly hazy, more or less rectangular patch of light. Raising the magnification to ×40 or ×80 brings out a bit more in the way of shape: I see this object as S-shaped at these higher powers, an impression reinforced in larger-aperture telescopes. None of the delicate, chaotic filamentary structure familiar from deep photographs and CCD images is evident in the unfiltered view in instruments even as large as 150 mm aperture. I find that use of an O III filter does little to alter the appearance of this object, but a UHC filter does apparently bring out more structure.

▲ An amateur CCD image of the Crab Nebula in Taurus, taken by Tony Pacey.

It is the remnant of a supernova which exploded in 1054.

M1 is known as the Crab Nebula from observations and drawings by Lord Rosse with a 36-inch (0.9 m) telescope during the early 1840s. The Crab has an overall diameter of $6' \times 4'$, corresponding at its distance of 6500 light years to an actual width of about 10 light years. It is estimated to be expanding at 6.5 million km/h.

To the eye, the Crab Nebula appears white. Its light comes from excited gas with a bluish component of synchrotron radiation produced by electrons traveling at high velocities in intense magnetic fields. The Crab Nebula is the remnant of a supernova seen in 1054, which was estimated to have had a magnitude of at least $-5$. This was a Type II event which left behind a pulsar spinning 30 times a second at its heart: at 16th magnitude, the pulsar is beyond reach of smaller telescopes.

M1, as the number suggests, was Messier's first entry in his catalog of cometary lookalikes. He came across the object on August 28, 1758, while searching for Halley's Comet (which was then about to make its first predicted return); indeed, it was his discovery of M1 that led him to compile the listing to avoid future confusion. The nebula had previously been recorded in 1731 on charts in *Uranographia Britannica* by the English amateur astronomer John Bevis (1695–1771).

The northern Milky Way around Taurus' horns contains a couple of rather faint supernova remnants which are accessible only with larger telescopes. S147 is an extended object with very low surface brightness centered roughly at RA 05h 05m, dec +28°, just south of 2nd-magnitude β Tau, the Bull's more northerly horn. S147 is the remnant of a supernova explosion that occurred 50,000 years ago.

IC 443 in Gemini is a more realistic prospect for medium-sized amateur telescopes, described by some observers as only slightly fainter than the Veil Nebula. IC 443 (RA 06h 17.1m, dec +22° 35′) has an angular size of $50' \times 40'$. Lying at a distance of about 5000 light years, it is perhaps 30,000 years old, and is sometimes referred to as the Medusa Nebula from its photographic appearance.

### The Veil Nebula

A well-known photographic object, the Veil Nebula (also known as the Cygnus Loop) is made up of the filamentary remnants of a star that exploded as a supernova 5000 years ago, according to measurements of its expansion rate by the Hubble Space Telescope. Material thrown out during the explosion continues to expand at a velocity of 600,000 km/h, creating a vast bubble in space. As it passes through the interstellar medium, the supernova shock wave compresses material ahead of it, leading to ionization and causing the faint glow of the filaments. A significant part of the emission comes from excited oxygen, and an O III or UHC filter will greatly enhance views of the Veil Nebula.

There are three principal regions of nebulosity here, centered at roughly RA 20h 50m, dec +32°: a couple of degrees SE of the 2nd-magnitude ε Cyg, the Swan's easterly wing.

### NGC 6992
Cygnus  RA 20h 56.4m dec +31° 43'  Map 6  Finder chart p.165
### NGC 6995
Cygnus  RA 20h 57.1m dec +31° 13'  Map 6  Finder chart p.165

The brightest section of the Veil Nebula complex lies at its eastern side, roughly marking the SE corner of a northward-pointing isosceles triangle with ε Cyg (the triangle's narrow tip) and the mag. +4.2 star 52 Cyg. NGC 6992, the northern part of this segment, is narrow

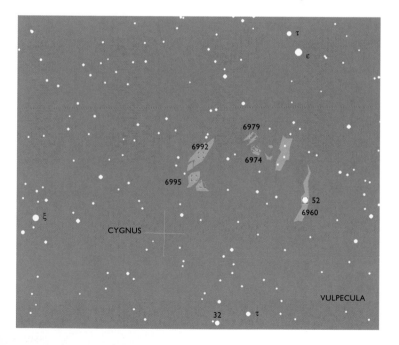

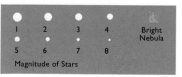

▲ Finder chart for the visually faint Veil Nebula, south of ε Cygni (the Swan's more easterly "wing"). The chart shows stars to limiting magnitude +8.5 and has an angular width of 8°.

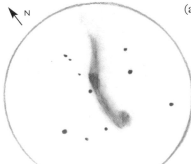

N

▲ *The eastern Veil Nebula, NGC 6992 and 6995, sketched by the author from the view in an 80 mm f/5 refractor at ×24 with an O III filter. Contrast is exaggerated here relative to the visual appearance.*

(about 8′–10′ wide) and curves SSE to NNW for the best part of a degree. On a night of good transparency, NGC 6992 can be seen in 10 × 50 binoculars as an elongated streak lacking internal structure.

The view is better in a small telescope. A minimum 80 mm aperture is recommended, and low magnification should be used – a wide-field view at ×20 or less is probably best – as the Veil has rather low surface brightness and contrast. With an instrument between 100 and 150 mm in aperture, higher magnification can be used. Such telescopes should also more clearly show the fainter, wedge-shaped NGC 6995 at the southern end of NGC 6992.

Some more definite hints of structure are seen through an O III or UHC filter, on larger instruments particularly. The eastern ("outer") edge of NGC 6992/6995 appears significantly brighter, for example, in the filtered view.

### NGC 6960
Cygnus   RA 20h 45.7m dec +30° 43′   Map 6   Finder chart p.165

The western side of the Veil Nebula is rather fainter than the eastern, and observation in small telescopes is made more difficult by the distracting relative brightness of the star 52 Cyg, whose glare may obscure NGC 6960. The star appears almost midway along the nebula's 1° length. Like NGC 6992, NGC 6960 is narrow, perhaps 5′–6′ wide. Beyond the reach of small binoculars, this is reasonably well seen in a 100 mm or larger telescope, and again low powers and an O III or UHC filter are recommended.

### Pickering's Triangular Wisp
Cygnus   RA 2h 48.5m dec +31° 09′   Map 6

### NGC 6974
Cygnus   RA 20h 50.8m dec +31° 52′   Map 6   Finder chart p.165

### NGC 6979
Cygnus   RA 20h 51.0m dec +32° 09′   Map 6   Finder chart p.165

The northern and central parts of the Veil are fainter still than the eastern and western extremities. Pickering's Triangular Wisp – not assigned an NGC number – lies about 1.5° NNE from 52 Cyg, and has dimensions of about 30′ × 15′, being narrower at the southern end. It is named for the American astronomer Edward C. Pickering (1846–1919). NGC 6974/6979 form a very faint, continuous patch aligned NW–SE, perhaps 8′–10′ long. These features are best seen in a large instruments with the aid of a nebula filter.

## The Vela Supernova Remnant

### NGC 2736

Vela   RA 09h 00.4m dec −45° 54′   Map 3

Part of a huge complex of faint nebulosity – the Gum Nebula, discovered photographically by the Australian astronomer Colin Gum (1924–60) – spanning parts of Puppis and Vela, the Vela Supernova Remnant is the brightest such object in southern skies. With an angular diameter of 8° (more than double that of the Veil in Cygnus), the Vela Supernova Remnant is rather faint. Amateur telescopes can show the brightest section, NGC 2736 – an isolated filament on the eastern side. Also known as the Pencil Nebula, this strand of nebulosity was first seen by John Herschel during his South African observing stint in 1835.

NGC 2736 is a long, thin (15′ × 5′) object about 80′ NNW of the 4th-magnitude star c Vel, and just west of an 8th-magnitude field star. Visually, it is a faint streak, comparable in visibility to parts of the Veil, and as with the Veil its appearance is greatly enhanced by use of an O III filter.

The Vela Supernova Remnant is at a distance of 1600 light years, and has an overall diameter of 200 light years. Material at its edge is traveling at a velocity of 600,000 km/h. At the center is the Vela Pulsar, a rapidly rotating neutron star – all that is left of the progenitor which exploded a thousand years ago according to recent estimates based on the pulsar's spin rate. Previous estimates suggested a more considerable age, in excess of 11,000 years.

# 10 · TAKING IT FURTHER

The two hundred or so objects described in the preceding chapters should give the newcomer to deep sky observing plenty of material for a good many nights under the stars: the tables at the end of this book indicate when each particular object is best placed for observation, and whatever the time of year there will be plenty to see. They are, of course, just a selection of some of the brightest from the celestial cornucopia: there are many, fainter targets to pursue. Accessing these leads to the next, almost inevitable step for the converted deep sky enthusiast – the acquisition of a bigger telescope, succumbing to what is sometimes described as aperture fever!

As discussed in Chapter 2, larger apertures allow the observer to see fainter objects. It's important to bear in mind that a telescope's performance will also be influenced by local observing conditions – there is little to be gained by the visual observer having a large telescope sited in a badly light-polluted location. Experienced observers tend to travel to places where the sky is as dark as possible, so their instruments have to be portable. For many, the best solution is to have a telescope designed to be readily assembled on arrival at the observing location, and dismantled into its component parts for transport home at the end of the session. Dobsonian-mounted reflectors are ideal for this – the rocker box, primary mirror assembly and secondary mirror/focusing mount can be separate components, connected by an open-tube frame of lightweight aluminum rods. When it comes to portable large-aperture instruments, most deep sky observers opt for the extra light grasp of a big-mirror Dobsonian over smaller instruments on driven equatorial mounts.

Most large spiral or elliptical galaxies are about the same actual size and have the same luminosity; so their magnitude depends mainly on their distance (apparent brightness decreases in proportion to the square of the distance, the so-called inverse square law). The more distant an object, the fainter it will appear. The most remote of the objects listed in the preceding chapters are galaxies 65 million light years away in the Virgo–Coma Cluster. Telescopes with apertures larger than 150 mm, the maximum considered in terms of access to the objects listed in this book, can reach fainter objects at greater distances. A good personal challenge for the observer is to see how far into the deep sky their instrument and skill can take them.

Many deep sky enthusiasts eventually become specialists, concentrating on a particular class of object. An observer specializing in globular clusters, for example, may spend a lifetime seeking out the 150 or so globulars associated with our Galaxy, a project which will require a trip to the opposite hemisphere since many targets will lie too far south or too far north to be observed from home.

Others may choose to pick off ever-fainter planetary nebulae, or follow variable nebulae over many years. Measurements of the slowly changing position angle and separation of the components in double star systems was once a very popular – and scientifically important (for determining orbital parameters) – pursuit for advanced amateur observers, and has regained some of its past importance as there are now many double stars which have gone unmeasured for decades.

### The Messier "Life List"

Deep sky observers – both specialists and generalists – rely heavily on charts, and on lists and catalogs of objects. A challenge that can be taken on even by novice observers is to see for oneself all 109 objects listed in the Messier catalog. This can take some time, and again geography can play its part. For example, among the four (at the time of writing) still missing from my own Messier "life list" after more than 30 years as an active observer is the low-contrast galaxy M83, which

▲ For construction of seriously large-aperture portable amateur telescopes,

Dobsonian designs are probably the best solution.

◀ *Sometimes known as the Splinter Galaxy, NGC 5866 in Draco is considered by some observers to be the "missing" M102.*

hugs the southern horizon from my home observing location. I am awaiting that rarest of phenomena – a truly haze-free, moonless, dark, English spring evening – for a chance at detecting it.

Working through the Messier list will take time, and "collecting" these objects should also be a good exercise in keeping records. As pointed out in Chapter 2, there is a world of difference between just looking at an object (being able to say you've seen it) and drawing it and making some detailed notes (being able to say you've *observed* it). It is worth keeping a separate logbook for your impressions and sketches of the Messier objects, which can be interesting to compare with those of other observers or with descriptions in the published literature. Several organizations, such as the Astronomical League in America, offer certificates to members who can provide detailed evidence that they have seen and recorded all the Messier objects.

### The Messier Marathon

Tracking down each of the Messier objects over the course of an observing career can be challenging enough for most amateur astronomers. But there are some, particularly in North America, and since the late 1970s, who have set out to spot all 109 of them in a single night's observing – the so-called Messier Marathon.

Some extend the full tally to 110 objects by including, as M102, the mag. +9.9 lenticular (S0) galaxy NGC 5866 (RA 15h 06.5m, dec +55° 46′) in Draco. Although this *may* have been recorded by Messier and/or Méchain in 1781, before William Herschel's 1788 discovery, its inclusion as the "missing" M102 – taken by most to have been a duplicate entry for M101 – remains contentious.

I have to admit to a certain ambivalence when it comes to the Messier Marathon. This may result in part from my roots in southwest Scotland, at latitude 56°N, from where M7 doesn't clear the horizon and several of the other southerly Messier objects culminate barely a couple of degrees up. Observers in, say, Sydney, find themselves similarly limited – for them, the Ursa Major M81/M82 galaxy pair is beyond reach, as are the open clusters M52 and M103 in Cassiopeia, while M101 barely gets above the horizon and even then is unlikely to be visible. In my opinion, an observer should have examined the Messier objects (or as many of them as possible) one at a time, for long enough to have gained a good impression, before taking on the Marathon with its necessary "quick look and move on" approach – to my mind at least, a celestial smash and grab!

Cynics might point out that a modern GOTO mounting can be programmed to slew the telescope to each of the Messier objects in turn. This would, of course, be regarded as cheating by those who see the Messier Marathon as a test of the observer's celestial navigation skills!

The Messier Marathon is possible only for a couple of weeks around the northern-hemisphere spring equinox, close to March 21 and from a latitude of about 40°N; at other times and from other latitudes at least some of the objects will be in conjunction behind the Sun. Before tackling the Marathon, it is necessary to plan the order in which the objects are to be taken.

Observers work their way eastward, starting in late evening twilight with the tricky galaxies M74 in Pisces (p.69), M77 in Cetus (p.69) and M33 in Triangulum (p.55) – all low in the western sky. The next leg,

▶ The last port of call in the usual Messier Marathon is the globular cluster M30 in Capricornus.

through Andromeda and on to the Taurus–Auriga–Orion region, can be more sedate. Some stages are more frantic than others – sweeping up the galaxies in the Virgo–Coma Cluster, or in the Sagittarius region, will seem more of a sprint than a marathon. Finally, as dawn begins to brighten, comes a big final challenge – the globular cluster M30 in Capricornus (p.84), low in the eastern twilight.

Observers in Northwest Europe find the Messier Marathon nigh-impossible to complete – M30 is just too low, and dawn comes very quickly at higher latitudes. More southerly locations give observers a better chance – perhaps explaining the Marathon's greater popularity in the United States. Amateur astronomers from the UK have succeeded by traveling to Spain or Portugal to undertake the Messier Marathon – a popular destination is the COAA observatory in the Algarve.

Success in completing the Marathon requires good planning and, of course, night-long clear skies – to some extent it's all a matter of luck, but at least there's always next year ...

### Beyond Messier

Many other unofficial or arbitrary lists of objects are available for observers to work through. The Astronomical League has on its web-site listings of the best clusters, galaxies, nebulae and double stars for binoculars and small telescopes, and they will keep the observer occu-pied for some time. For protracted "object bagging" there are more extensive challenges such as the "Herschel 400" – a set of 400 NGC objects originally recorded by William Herschel in the late eighteenth century, compiled by the Ancient City Astronomy Club in St Augustine, Florida in the USA.

The Royal Astronomical Society of Canada (RASC) provides lists of 45 challenging targets and a season-by-season selection of the 110 finest NGC objects in its annual *Observer's Handbook*.

The original NGC contained a number of poorly defined open clusters. Deep sky observers have been encouraged to re-examine some of these using amateur equipment as part of a "cluster recov-ery" project initiated by the UK-based observers' magazine *The Astronomer* in the late 1970s, and continued more recently by the Webb Society (see below).

### Observing Groups

While a lot of the attraction of deep sky observing comes from person-al satisfaction, most amateur astronomers will be keen to share and compare their results with others. Local astronomical clubs and soci-eties are found worldwide, and their regular meetings provide an information exchange and a valuable source of advice.

National and international bodies also play an important role in helping observers and collecting together reports. The Astronomical League has already been mentioned as a useful source of observing lists. The British Astronomical Association, reviving a tradition stretching back to its early days in the late nineteenth century when it had observing sections dedicated to areas such as star colors and double star measurements, established a Deep Sky Section in the 1980s to cater for the increasing interest among its members for this sort of observing. Similarly, the RASC and Astronomical Society of South Africa have deep sky groups.

Internationally, the Webb Society is the premier organization for deep sky enthusiasts. Founded in 1967, it is named for Thomas William Webb, the clergyman whose *Celestial Objects for Common Telescopes* was one of the earliest deep sky handbooks, providing a guide to numerous interesting objects. Webb observed from Hardwicke in Herefordshire, near the English/Welsh border, and while his descriptions may seem quaint by twenty-first-century standards, his two-volume book still makes interesting reading.

Maintaining the tradition of Webb and other nineteenth-century observers, the Webb Society encourages its members to examine deep sky objects, and collates their observing reports. Over the years, the Webb Society's quarterly journal *The Deep Sky Observer* has carried articles describing some very challenging objects and observing projects, and has set the standards in detailed reporting. Specialized sections deal with double stars, nebulae and clusters, galaxies, and the southern sky. The Webb Society has produced a number of useful observing handbooks covering the various classes of deep sky object.

▲ *The Rev. T. W. Webb, a popularizer of observational astronomy in the nineteenth century.*

While amateur astronomy is often perceived as a strictly solitary pursuit, there can be little doubt that the exchange of observations and experience with others can be of considerable benefit – as demonstrated by the enduring popularity of star parties, where many observers gather beneath dark skies, and other events where like-minded individuals can get together. Membership of a local club, one's national amateur body or a group with international scope like the Webb Society will certainly help to make you a better observer. And if taking it all further leads you into adventures in photography or CCD imaging, the wisdom and experience of others – who will have already encountered the pitfalls – will prove invaluable.

Above all, observing the deep sky and hunting down some of its more elusive targets should be seen as a challenge to be enjoyed. There are few experiences to compare with the thrill of finally tracking down an object whose attenuated light has spent millions of years traveling through space to be gathered by your retina. Observing the deep sky can be frustrated by poor weather or light pollution, but patience, perseverance and skill gained through experience will be rewarded in the long run.

# 11 · A BRIEF HISTORY OF DEEP SKY OBSERVING

**M**any of the prominent, naked-eye deep sky objects have been known since prehistoric times, among them the Hyades and the Pleiades, and the Magellanic Clouds. Other, less obvious objects were certainly known in antiquity. In the catalog compiled by the Greek astronomer Hipparchus in the second century BC, both the Praesepe in Cancer and the Double Cluster in Perseus are mentioned as "clouds." Based substantially on Hipparchus' work, Ptolemy of Alexandria's *Almagest* (second century AD) also includes the bright Scorpius star cluster later known as M7.

The *Almagest* remained the principal stellar listing used by astronomers for several centuries. The Persian astronomer al-Sufi compiled a later listing, the *Book of the Fixed Stars* in 964, in which the Andromeda Galaxy is mentioned as a "little cloud." Al- Sufi also noted a cluster which may have been IC 2391, in Vela.

## Early Telescopic Observers

Beyond the small selection of obviously non–stellar naked-eye objects, the unveiling of the deep sky had to await the development of the telescope and its application to astronomy by Galileo and others from 1609 onward. Galileo's early observations included the discovery that Praesepe was in fact made up of numerous stars beyond the limit of naked-eye resolution, while his telescopic explorations also showed the Milky Way's hazy band to be the combined light of countless stars.

Early refractors suffered badly from chromatic aberration, and Galileo's first telescopes would compare poorly even with some of today's mass-produced high-street models. Before the discovery that different types of glass could be combined in achromatic objective lenses, telescope-makers sought to reduce the image-degrading effects of chromatic aberration by constructing telescopes with very long focal length. These so-called aerial telescopes could have objectives with focal lengths of many meters, and were unwieldy to say the least. Consequently, the deep sky received little attention in the early seventeenth century, as astronomers concentrated on more easily observed targets such as the planets.

Noted for his observations of Saturn's rings, and his drawing of Mars showing the dark feature known today as Syrtis Major, the Dutch astronomer Christiaan Huygens (1629–95) did turn a long-focus telescope on the Orion Nebula, in 1656, and was the first to comment on the dark bay of the Fish's Mouth. At Danzig (now Gdansk), Huygens' contemporary Johannes Hevelius also used an aerial telescope for some of his observational work, and recorded 16 "nebulous" objects including the pair of stars in Ursa Major later cataloged by Messier as M40.

Mostly, deep sky objects found during that time were swept up by accident as observers studied what they regarded as more interesting targets. The globular cluster M22 in Sagittarius was apparently found by the German astronomer Abraham Ihle (1627–99) while he was observing Saturn. Gottfried Kirch at Berlin found the rich open cluster we know as M11 in 1681. On May 5, 1702, Kirch and his wife Maria (1670–1720) discovered the Serpens globular cluster M5 while they were observing a comet.

More systematic telescopic charting of the sky in the later seventeenth century led to further discoveries. Edmond Halley, who would gain fame for calculating the orbit of the comet that now bears his name, and become the second Astronomer Royal, undertook an expedition to St Helena to chart the southern stars. From there he recorded six nebulous objects (now known as M11, M22, M31, M42, M13 and Omega Centauri; Halley is credited with discovery of the last two of these).

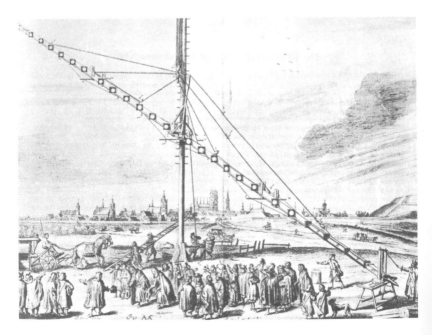

▲ Seventeenth-century astronomers observing with unwieldy aerial telescopes tended to concentrate on brighter, easily located targets like the planets.

## Discoveries in the Early Eighteenth Century

If ever there was a golden age of deep sky discovery, it was surely in the eighteenth century. It came about not so much from a desire to study the "nebulae" themselves, but from the search for comets, which was prompted in part by the new understanding of the Solar System and orbital dynamics. Again, in hindsight, this should not be surprising – the nature of "nebulae" (a catch-all description that at the time took in galaxies, gas and dust clouds, and unresolved star clusters) as objects extending the scale of the Universe well beyond the neighborhood of the Solar System was far from being comprehended. Comet-hunters often came across nebulous objects as they swept the sky, and some began to catalog them.

The best known of these catalogs is the one produced by Messier. An earlier listing was compiled by the Swiss astronomer Philippe Chésaux in 1746. It contained 21 objects, several of which later featured in Messier's list, appearing as M4, M16, M17, M25 and M35. Presented in a paper to the French Academy of Sciences, de Chésaux's catalog never enjoyed a wide circulation.

### Lacaille and the Southern Sky

Surveys of the southern sky continued in the eighteenth century, too. The Frenchman Nicolas de Lacaille spent two years at the Cape of Good Hope (now part of South Africa) from 1751 to 1753, during which time he measured the positions of 9776 stars, published in 1763 in his *Coelum australe stelliferum*. Although he had only very small telescopes at his disposal, he managed to observe and catalog 42 nebulous objects – 14 of each of three categories (nebulae, nebulous star clusters and nebulous stars). Among them were Omega Centauri, 47 Tucanae, the Kappa Crucis Cluster, and the objects later known as M22 and M55. A significant Lacaille discovery was M83, the Hydra galaxy later identified as the first object to have been found beyond the Local Group, to which our Milky Way belongs. Lacaille's catalog was presented to the French Academy of Sciences in 1755, and Messier appended it to later versions of his listings.

### Messier and his Catalog

Most famous of the eighteenth-century comet-hunters is, of course, the French astronomer Charles Messier. Observing from Paris in the 1750s, Messier was engaged in the search for Halley's Comet on its first predicted return in 1758–9. He made an independent recovery of the comet in January 1759, some weeks after the initial sighting by Johann Palitzsch (1723–88) from Saxony. During the search, Messier came across the Crab Nebula in Taurus, which became the first entry in his

catalog of comet-like objects. His comet searches were very successful –
he is acknowledged as sole discoverer of at least 14, and co-discoverer
of many more – and it is a little ironic that he is best remembered not
for these, but for his list of objects that might be mistaken for comets.

The period from 1764 onward was particularly productive for
Messier in terms of object discoveries, leading to the first published
version of his catalog in 1774, listing 45 objects including several found
by other observers. A further 23 objects were added in a 1780 supple-
ment, which appeared in the 1783 edition of the French almanac
*Connaissance des Temps*.

During the 1780s, Messier was joined at Paris by Pierre Méchain,
another comet-hunter. Although he competed with Messier for discoveries, Méchain was by all accounts a friendly rival, and when he
turned up new nebulous objects he communicated his findings to
Messier. Once he had determined accurate positions for them,
Messier added Méchain's objects to later versions of his catalog. By
April 1781, the combined efforts of Messier and Méchain (the former continuing to make discoveries in his own right) had raised the
catalog's total to 104 objects.

The final extension of the accepted Messier listing
to 110 objects (albeit that M102 is a duplication
of M101) results from twentieth-century
research into papers left by Messier and
Méchain. For instance, M104 (the
Sombrero Galaxy in Virgo) was added
by the French popularizer of astronomy Camille Flammarion (1842–1925)
in 1921.

Contemporary with Messier and
Méchain was the German Johannes
Bode, who from the Berlin Observatory
discovered a number of objects including the Ursa Major galaxies included in
Messier's catalog as M81 and M82, and
the Coma Berenices globular cluster M53.
Bode published his catalog of 77 nebulous
objects ("A Complete Catalogue of Hitherto
Observed Nebulous Stars and Clusters") in
the *Astronomische Jahrbuch* for 1799.

▲ *The French comet-
hunter Charles Messier
compiled what is now the
best-known catalog of
nebulous objects used by
deep sky observers.*

## William Herschel

Among the most important of Messier's con-
temporaries was William Herschel. Born in

Hanover, Herschel moved to England in 1757 to pursue a career as a musician. Employed as an organist in Bath, Herschel was able to pursue his astronomical interests from 1773 onward, initially using home-built reflectors of 6-inch (150 mm) and 8-inch (200 mm) aperture. His home at 19 King Street, Bath, is now preserved as the Herschel House Museum.

Herschel made a systematic survey of the sky in strips, logging numbers and positions of stars and other objects. He was assisted in this task by his sister Caroline Herschel (1750–1858) – later (from 1782) to prove herself a skilled observer, discovering eight comets and several deep sky objects including the galaxies NGC 253 and NGC 891.

A key moment in William Herschel's observing career came on March 13, 1781, when he discovered the planet Uranus (then against the stars of Gemini). This led to a knighthood and royal patronage, from King George III, and Sir William re-located to Datchet, near Windsor, where he constructed and used large telescopes. His 48-inch (1.2-meter) reflector was, in the late 1780s, the largest telescope in the world.

Part of Herschel's aim was to establish the distribution of stars in space. From his studies of the stellar densities of countless starfields, he surmised that the Sun was located in a flat, roughly disk-shaped collection of stars, though his model was still some way from being the fully realized Galaxy familiar to modern-day astronomers. William Herschel recorded and measured over a thousand double stars, and also cataloged more than 2000 nebulae and clusters.

### The Nineteenth Century

Herschel's work was continued and extended by his son John, who discovered more nebulae and clusters from England, and also undertook a five-year expedition to the Cape of Good Hope, beginning in 1834, to survey the southern sky using one of his father's large telescopes. From southern Africa, John Herschel added 1200 double stars and 1700 nebulae and clusters to William's listings. John collated his own and

▲ William Herschel conducted a systematic survey of the sky, cataloguing thousands of nebulae and star clusters. He laid the basis for the NGC, still in use today.

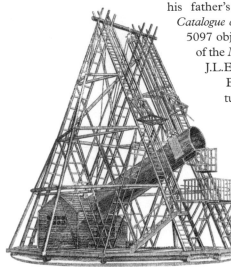

▲ *Herschel's 48-inch reflector was, in the late 1780s, the largest telescope in the world.*

his father's observations into the *General Catalogue of Clusters and Nebulae* (a listing of 5097 objects), which later formed the basis of the *New General Catalogue* published by J.L.E. Dreyer in 1887.

By the middle of the nineteenth century, the telescopic equipment available to the wealthier amateur astronomers of the day had improved considerably. Among the "gentleman amateurs" who made significant contributions to astronomical research at this time were observers such as William Rutter Dawes (1799–1868), an English physician and clergyman who measured double stars using a number of telescopes, including a 200 mm refractor, and established the Dawes limit as a measure of the resolving power of a telescope (p.134).

Other notable observers of this period included Admiral Smyth, who in 1830 established an observatory at Bedford equipped with a 142 mm aperture refractor. Smyth observed hundreds of double stars, clusters, galaxies and nebulae, compiling his *Cycle of Celestial Objects* in 1844 – the first guidebook to the deep sky for an amateur readership. A flavor of Smyth's work is given by his description of M92 in Hercules: "A globular cluster of minute stars, preceding the right leg of Hercules. The object is large, bright and resolvable, with a very luminous center; and under the best views, has irregular streamy edges."

Another amateur who sought to popularize observing around this time was the Rev. T.W. Webb, an early advocate of silvered-mirror Newtonian reflectors which gave access to larger apertures: most amateurs up to this time used quite small-aperture refractors.

### The Leviathan of Parsonstown

The finest example of the application of large reflectors to studying the visual deep sky can be found in the endeavors of William Parsons, the Third Earl of Rosse, commonly known as Lord Rosse. His largest telescope – and the world's largest until the completion of the 100-inch (2.5-meter) Hooker Telescope at Mount Wilson, California, in 1917 –

▶ *Lord Rosse's "Leviathan of Parsonstown," a 72-inch reflector, was the world's largest telescope in the late nineteenth century. It was restored at the end of the twentieth century.*

was a 72-inch (1.8-meter) reflector, completed in 1845. Sited at Birr Castle in County Offaly, Ireland, it was nicknamed the Leviathan of Parsonstown. This huge telescope had a mirror cast in speculum metal (a highly reflective amalgam of copper and tin) and a long tube supported between two massive masonry walls. Movement in azimuth was limited, so objects could be followed only for an hour or so each night when they were close to the meridian. It was with this telescope that Rosse was able to make out the spiral structure of M51, which became known thereafter as the Whirlpool Galaxy. Rosse used the Leviathan to study a number of other galaxies which he found to have a similar spiral form. His drawings of fine structure in M1 led to its acquisition of the Crab Nebula tag.

Birr Castle was for a time the focal point for astronomical observation in Ireland, and the Fourth Earl continued some of his father's work into the 1870s. J.L.E. Dreyer, later director of Armagh Observatory and compiler of the NGC, worked there from 1874 to 1878.

The 72-inch Leviathan was dismantled in 1916. Restoration work using the original tube and brickwork was completed in the late 1990s, and visitors to Birr Castle today can get some impression of what the telescope was like in its heyday. Having stood at the observing platform, more than ten meters above ground level with a precipitous drop below, one can only admire the courage of the observers who used this telescope in the dark of the Irish night!

## The Rise of Astrophysics

Technological developments during the nineteenth century gradually brought the cutting edge in deep sky study closer to the realm of professional astronomers. Improvements in photography and its application to astronomical imaging from the 1880s allowed professional astronomers to probe to fainter magnitudes than the eye could reach, opening the door to fundamental discoveries in the field of cosmology.

To many historians of science, astrophysics was born once astronomers were able to analyze starlight in detail by spectroscopy. The pioneering work of the amateur astronomer William Huggins at Tulse Hill, South London was mentioned earlier in connection with planetary nebulae. Working with William Miller (1817–70), professor of chemistry at King's College, London, Huggins made the first analysis of stellar spectra based on a comparison with emissions from familiar chemical elements studied in terrestrial laboratories. In addition to discovering the O III emission from planetary nebulae, Huggins' work demonstrated that the spectra of the "spiral nebulae," which we now know to be distant galaxies, were of the same continuous nature as those of stars.

By the early twentieth century, progress in equipment and technology had finally moved the leading edge of discovery away from amateur astronomers and their relatively limited equipment and into the hands

▲ The William Herschel Telescope on La Palma is one of many professional instruments probing the deeper Universe. Amateur astronomers can no longer compete scientifically with their professional counterparts, but there is much pleasure to be derived from simple visual observing with small telescopes.

of professional astronomers. The construction of very large professional telescopes, including the 100-inch Hooker Telescope and the 200-inch (5-meter) Hale Telescope at Mount Palomar, gave access to ever-fainter objects and remote galaxies. Edwin Hubble proved that the spiral "nebulae" really were extremely remote equivalents of our own Milky Way Galaxy, and from the study of the spectra of distant galaxies astronomers came to appreciate the Universe's ongoing expansion.

## *Amateur Observers Still Matter*

An interesting final twist to the technological advances of the twentieth century and into the twenty-first has been the adoption of professional-standard equipment by amateur astronomers. Using commercially available CCD cameras, today's advanced amateurs are able to obtain images of deep sky objects which compare favorably with professional work of only a couple of decades' vintage. This relatively new electronic imaging technology also allows those with the dedication to pursue deep sky observing as serious research again – in terms of making supernova discoveries, early alerts to which are of value to professionals. The ability to contribute has been restored!

For most of us, however, deep sky observing simply offers the chance to see for ourselves some of the objects whose study by professional astronomers has opened up our understanding of our place in the Universe – and there is absolutely nothing wrong with that! Above all, remember that astronomy can be pursued simply for the pleasure it brings, and that any scientifically significant results that follow from amateur observations can fairly be considered a bonus.

# STAR CHARTS

The accompanying wide-field charts show stars to limiting magnitude +5, representing the appearance of the constellations from a reasonably dark observing site. North circumpolar stars are shown on p.185, south circumpolar stars on p.192. The equatorial regions of the celestial sphere are shown on pages 186–191. Right ascension is shown along the top and bottom of the charts, declination on the left and right edges. Lighter shading indicates the rough outline of the Milky Way.

These low-resolution charts provide a general guide to positions for the objects discussed in the book. Closer inspection shows open clusters to be commonest along the plane of the Milky Way (our Galaxy's spiral arms), while globular cluster are mostly found in the constellations surrounding the Galactic center in Sagittarius. Distant galaxies are more abundant in those parts of the sky like Virgo, Coma Berenices and Leo which lie away from the Milky Way and where we have a clear view into intergalactic space.

The heading for each object description gives the map on which that object can be found. A list of which constellations appear on which maps is given on the first page of the index.

| Greek alphabet | | | | | |
|---|---|---|---|---|---|
| α | alpha | ι | iota | ρ | rho |
| β | beta | κ | kappa | σ | sigma |
| γ | gamma | λ | lambda | τ | tau |
| δ | delta | μ | mu | υ | upsilon |
| ε | epsilon | ν | nu | φ | phi |
| ζ | zeta | ξ | xi | χ | chi |
| η | eta | o | omicron | ψ | psi |
| θ | theta | π | pi | ω | omega |

Map 1

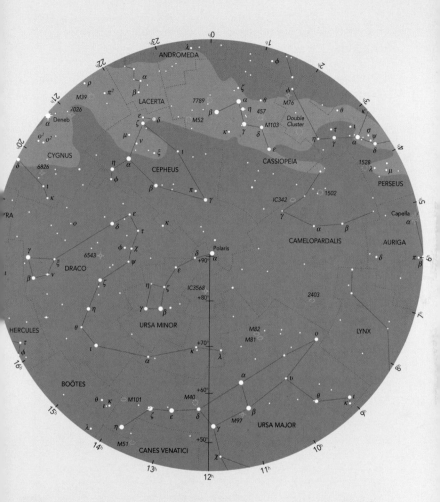

## Key to star charts

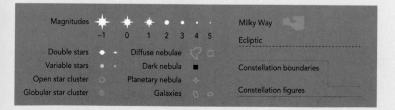

| Magnitudes | | Milky Way |
| --- | --- | --- |
| −1  0  1  2  3  4  5 | | |
| Double stars | Diffuse nebulae | Ecliptic |
| Variable stars | Dark nebula | Constellation boundaries |
| Open star cluster | Planetary nebula | |
| Globular star cluster | Galaxies | Constellation figures |

## Map 2

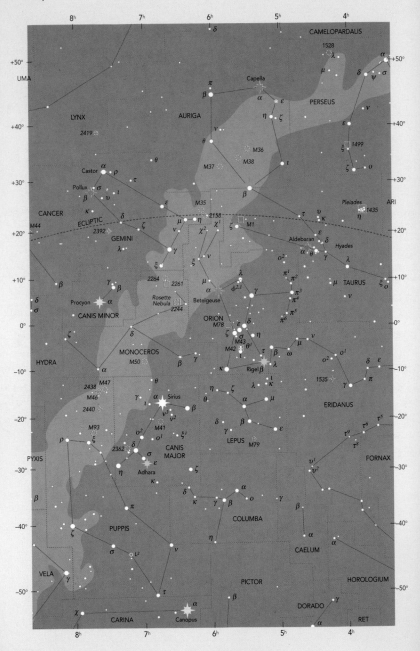

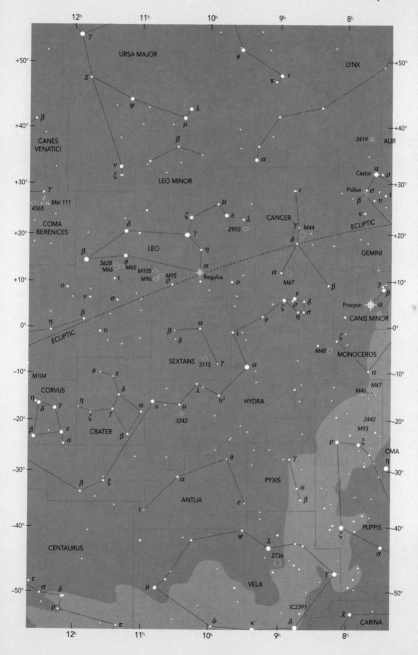

## Map 4

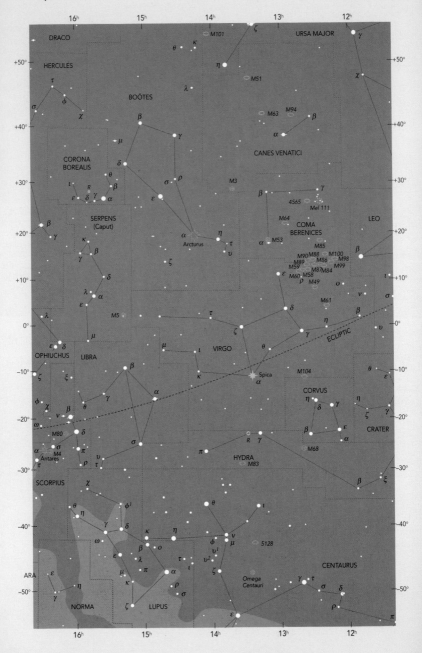

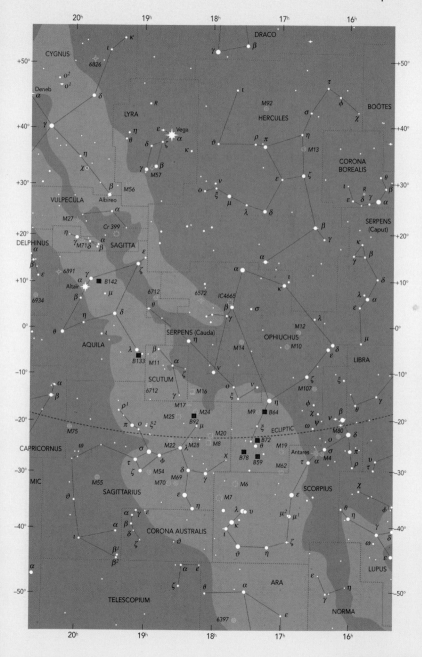

## Map 6

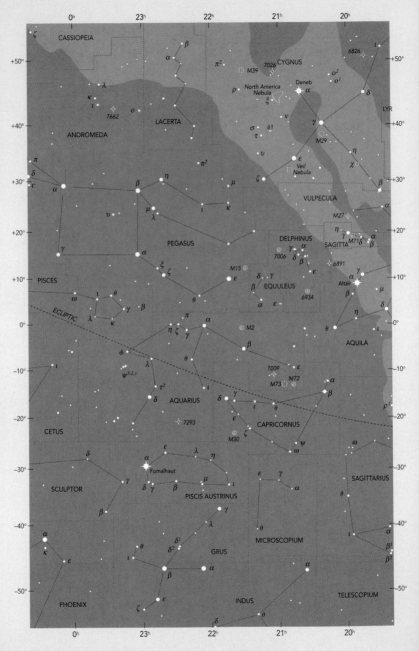

Map 7

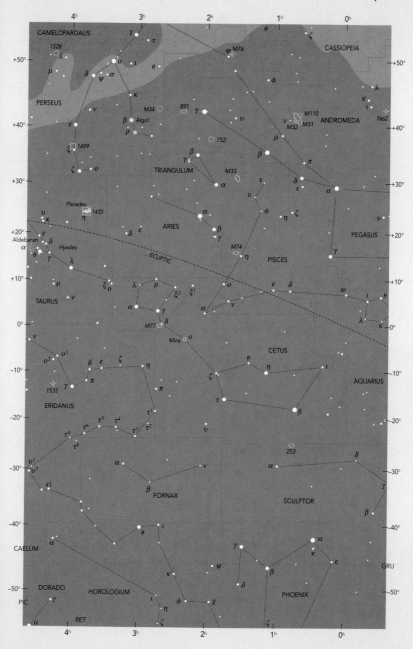

# Map 8

# DEEP SKY LISTINGS

## Objects by Constellation

Summary information for all the objects described in this book is provided in the tables which follow. Listings for objects within each constellation are provided in order of right ascension.

| Designation/Name | RA h | m | Dec ° | ′ | Mag. | Type | Page |
|---|---|---|---|---|---|---|---|
| **Objects by Constellation** | | | | | | | |
| **Andromeda (And)** | | | | | | | |
| NGC 7662 Blue Snowball | 23 | 25.9 | +42 | 33 | 8.3 | Planetary Nebula | 156 |
| M110 NGC 205 | 00 | 40.4 | +41 | 41 | 8.1 | Galaxy | 53 |
| M31 NGC 224 Andromeda Galaxy | 00 | 42.7 | +41 | 16 | 3.4 | Galaxy | 53 |
| M32 NGC 221 | 00 | 42.7 | +40 | 52 | 8.1 | Galaxy | 53 |
| NGC 752 | 01 | 57.8 | +37 | 41 | 5.7 | Open Cluster | 130 |
| γ And Almach | 02 | 03.9 | +42 | 20 | | Double Star | 140 |
| NGC 891 | 02 | 22.6 | +42 | 21 | 9.9 | Galaxy | 71 |
| **Aquarius (Aqr)** | | | | | | | |
| M72 NGC 6981 | 20 | 53.5 | −12 | 32 | 9.3 | Globular Cluster | 83 |
| M73 NGC 6994 | 20 | 59.0 | −12 | 58 | | Asterism | 131 |
| NGC 7009 Saturn Nebula | 21 | 04.2 | −11 | 22 | 8.3 | Planetary Nebula | 157 |
| M2 NGC 7089 | 21 | 33.5 | −00 | 49 | 6.4 | Globular Cluster | 83 |
| ζ Aqr | 22 | 28.8 | −00 | 01 | | Double Star | 145 |
| NGC 7293 Helix Nebula | 22 | 29.6 | −20 | 48 | 7.3 | Planetary Nebula | 157 |
| **Aquila (Aql)** | | | | | | | |
| B133 | 19 | 06.1 | −06 | 50 | | Dark Nebula | 110 |
| B142 | 19 | 40.7 | +10 | 57 | | Dark Nebula | 110 |
| **Ara (Ara)** | | | | | | | |
| NGC 6397 Ara | 17 | 40.7 | −53 | 40 | 5.9 | Globular Cluster | 81 |
| **Aries (Ari)** | | | | | | | |
| γ Ari | 01 | 53.5 | +19 | 18 | | Double Star | 138 |
| ε Ari | 02 | 59.2 | +21 | 20 | | Double Star | 145 |
| **Auriga (Aur)** | | | | | | | |
| M38 NGC 1912 | 05 | 28.7 | +39 | 50 | 6.4 | Open Cluster | 120 |
| M36 NGC 1960 | 05 | 36.1 | +34 | 08 | 6.0 | Open Cluster | 120 |
| M37 NGC 2099 | 05 | 52.4 | +32 | 33 | 5.6 | Open Cluster | 121 |
| **Boötes (Boo)** | | | | | | | |
| ε Boo Izar | 14 | 45.0 | +27 | 04 | | Double Star | 144 |
| **Camelopardalis (Cam)** | | | | | | | |
| IC 342 | 03 | 46.8 | +68 | 06 | 8.4 | Galaxy | 57 |
| Kemble's Cascade | 04 | 10 | +63 | | | Asterism | 132 |
| NGC 2403 | 07 | 36.9 | +65 | 36 | 8.5 | Galaxy | 57 |
| IC 3568 | 12 | 32.9 | +82 | 33 | 10.6 | Planetary Nebula | 159 |
| **Cancer (Cnc)** | | | | | | | |
| M44 NGC 2632 Praesepe | 08 | 40.1 | +19 | 59 | 3.1 | Open Cluster | 117 |
| ι Cnc | 08 | 46.7 | +28 | 46 | | Double Star | 140 |
| M67 NGC 2682 | 08 | 50.4 | +11 | 49 | 6.9 | Open Cluster | 118 |

| Designation/Name | RA h | m | Dec ° | ′ | Mag. | Type | Page |
|---|---|---|---|---|---|---|---|
| **Canes Venatici (CVn)** | | | | | | | |
| M94 NGC 4736 | 12 | 50.9 | +41 | 07 | 8.2 | Galaxy | 62 |
| α CVn (Cor Caroli) | 12 | 56.0 | +38 | 19 | | Double Star | 137 |
| M63 NGC 5055 Sunflower Galaxy | 13 | 15.8 | +42 | 02 | 8.6 | Galaxy | 61 |
| M51 NGC 5194 Whirlpool Galaxy | 13 | 29.9 | +47 | 12 | 8.4 | Galaxy | 60 |
| M3 NGC 5272 | 13 | 42.2 | +28 | 23 | 5.9 | Globular Cluster | 81 |
| **Canis Major (CMa)** | | | | | | | |
| M41 NGC 2287 | 06 | 47.0 | −20 | 44 | 4.5 | Open Cluster | 123 |
| NGC 2362 Tau Canis Majoris Cluster | 07 | 18.8 | −26 | 47 | 4.1 | Open Cluster | 124 |
| **Capricornus (Cap)** | | | | | | | |
| β Cap | 20 | 21.0 | −14 | 47 | | Double Star | 139 |
| M30 NGC 7099 Cap | 21 | 40.4 | −23 | 11 | 7.3 | Globular Cluster | 84 |
| **Carina (Car)** | | | | | | | |
| IC 2602 Southern Pleiades | 10 | 43.2 | −64 | 24 | 1.9 | Open Cluster | 117 |
| NGC 3372 Eta Carinae Nebula | 10 | 43.8 | −59 | 52 | 1.0 | Emission Nebula | 98 |
| **Cassiopeia (Cas)** | | | | | | | |
| M52 NGC 7654 | 23 | 24.2 | +61 | 35 | 6.3 | Open Cluster | 128 |
| NGC 457 | 01 | 19.1 | +58 | 20 | 6.4 | Open Cluster | 129 |
| M103 NGC 581 | 01 | 33.2 | +60 | 52 | 7.4 | Open Cluster | 129 |
| **Centaurus (Cen)** | | | | | | | |
| NGC 5139 Omega Centauri | 13 | 26.8 | −47 | 29 | 3.5 | Globular Cluster | 76 |
| NGC 5128 | 13 | 25.5 | −43 | 01 | 6.7 | Galaxy | 58 |
| α Cen (Rigil Kentaurus) | 14 | 39.6 | −60 | 50 | | Double Star | 137 |
| **Cetus (Cet)** | | | | | | | |
| M77 NGC 1068 | 02 | 42.7 | −00 | 01 | 8.9 | Galaxy | 69 |
| **Coma Berenices (Com)** | | | | | | | |
| M98 NGC 4192 | 12 | 13.8 | +14 | 54 | 10.1 | Galaxy | 64 |
| M99 NGC 4254 | 12 | 18.8 | +14 | 25 | 9.9 | Galaxy | 64 |
| M100 NGC 4321 | 12 | 22.6 | +15 | 47 | 9.3 | Galaxy | 64 |
| Mel 111 | 12 | 25 | +26 | | 1.8 | Open Cluster | 116 |
| M85 NGC 4382 | 12 | 25.4 | +18 | 11 | 9.1 | Galaxy | 64 |
| M88 NGC 4501 | 12 | 32.0 | +14 | 25 | 9.6 | Galaxy | 64 |
| M91 NGC 4548 | 12 | 35.4 | +14 | 30 | 10.2 | Galaxy | 64 |
| NGC 4565 | 12 | 36.3 | +25 | 59 | 9.6 | Galaxy | 71 |
| M64 NGC 4826 Black Eye Galaxy | 12 | 56.7 | +21 | 41 | 8.5 | Galaxy | 62 |
| M53 NGC 5024 | 13 | 12.9 | +18 | 10 | 7.5 | Globular Cluster | 82 |
| **Crux (Cru)** | | | | | | | |
| α Cru (Acrux) | 12 | 26.6 | −63 | 06 | | Double Star | 145 |
| NGC 4755 Jewel Box | 15 | 53.6 | −60 | 20 | 4.2 | Open Cluster | 126 |
| **Cygnus (Cyg)** | | | | | | | |
| β Cyg (Albireo) | 19 | 30.7 | +27 | 58 | | Double Star | 139 |
| NGC 6826 Blinking Planetary | 19 | 44.8 | +50 | 31 | 8.8 | Planetary Nebula | 155 |
| o¹ Cyg | 20 | 13.6 | +46 | 44 | | Multiple Star | 140 |
| 61 Cyg | 20 | 13.6 | +38 | 45 | | Double Star | 138 |
| M29 NGC 6913 | 20 | 23.9 | +38 | 32 | 6.6 | Open Cluster | 130 |
| Veil Nebula | 20 | 50 | +32 | | | Supernova Remnant | 164 |
| NGC 7000 North America Nebula | 20 | 58.8 | +44 | 20 | 5? | Emission Nebula | 104 |

| Designation/Name | RA h | m | Dec ° | ′ | Mag. | Type | Page |
|---|---|---|---|---|---|---|---|
| NGC 7026 | 21 | 06.3 | +47 | 51 | 10.9 | Planetary Nebula | 156 |
| M39 NGC 7092 | 21 | 32.2 | +48 | 26 | 5.0 | Open Cluster | 130 |
| **Delphinus (Del)** | | | | | | | |
| NGC 6891 | 20 | 15.2 | +12 | 42 | 10.5 | Planetary Nebula | 159 |
| NGC 6934 | 20 | 34.2 | +07 | 24 | 8.7 | Globular Cluster | 91 |
| γ Del | 20 | 46.7 | +16 | 07 | | Double Star | 141 |
| NGC 7006 | 21 | 01.4 | +16 | 12 | 10.5 | Globular Cluster | 91 |
| **Dorado (Dor)** | | | | | | | |
| Large Magellanic Cloud | 05 | 23.6 | −69 | 45 | 0.1 | Galaxy | 51 |
| NGC 2070 Tarantula Nebula | 05 | 38.7 | −69 | 06 | 8.0(?) | Emission Nebula | 103 |
| **Draco (Dra)** | | | | | | | |
| ν Dra | 17 | 32.2 | +55 | 10 | | Double Star | 139 |
| NGC 6543 Cat's Eye Nebula | 17 | 58.6 | +66 | 38 | 8.1 | Planetary Nebula | 153 |
| **Eridanus (Eri)** | | | | | | | |
| NGC 1535 | 04 | 14.2 | −12 | 44 | 9.6 | Planetary Nebula | 158 |
| o² Eri | 04 | 15.3 | −07 | 39 | | Double Star | 137 |
| **Gemini (Gem)** | | | | | | | |
| NGC 2158 | 06 | 07.5 | +23 | 18 | 8.6 | Open Cluster | 122 |
| M35 NGC 2168 | 06 | 08.9 | +24 | 20 | 5.1 | Open Cluster | 122 |
| NGC 2392 Eskimo Nebula | 07 | 29.2 | +20 | 55 | 9.2 | Planetary Nebula | 151 |
| α Gem (Castor) | 07 | 34.6 | +31 | 53 | | Double Star | 143 |
| **Hercules (Her)** | | | | | | | |
| M13 NGC 6205 | 16 | 41.7 | +36 | 28 | 5.7 | Globular Cluster | 79 |
| α Her (Rasalgethi) | 17 | 14.6 | +14 | 23 | | Double Star | 141 |
| M92 NGC 6341 | 17 | 17.1 | +43 | 08 | 6.5 | Globular Cluster | 80 |
| **Hydra (Hya)** | | | | | | | |
| M48 NGC 2458 | 08 | 13.8 | −05 | 48 | 5.8 | Open Cluster | 125 |
| NGC 3242 Ghost of Jupiter | 10 | 24.8 | −18 | 38 | 7.8 | Planetary Nebula | 154 |
| M68 NGC 4590 | 12 | 39.5 | −26 | 45 | 7.7 | Globular Cluster | 89 |
| M83 NGC 5236 | 13 | 37.0 | −29 | 52 | 7.6 | Galaxy | 70 |
| **Leo (Leo)** | | | | | | | |
| NGC 2903 | 09 | 32.2 | +21 | 30 | 9.0 | Galaxy | 60 |
| γ Leo (Algieba) | 10 | 20.0 | +19 | 51 | | Double Star | 144 |
| M95 NGC 3351 | 10 | 44.0 | +11 | 42 | 9.7 | Galaxy | 59 |
| M96 NGC 3368 | 10 | 46.8 | +10 | 49 | 9.2 | Galaxy | 59 |
| M105 NGC 3379 | 10 | 47.8 | +12 | 35 | 9.3 | Galaxy | 59 |
| M65 NGC 3623 | 11 | 18.9 | +13 | 05 | 9.3 | Galaxy | 58 |
| M66 NGC 3627 | 11 | 20.2 | +12 | 59 | 8.9 | Galaxy | 58 |
| NGC 3628 | 11 | 20.3 | +13 | 36 | 9.5 | Galaxy | 58 |
| **Lepus (Lep)** | | | | | | | |
| M79 NGC 1904 | 05 | 24.5 | −24 | 33 | 7.8 | Globular Cluster | 87 |
| **Lynx (Lyn)** | | | | | | | |
| NGC 2419 Intergalactic Tramp | 07 | 38.1 | +38 | 53 | 10.3 | Globular Cluster | 93 |
| **Lyra (Lyr)** | | | | | | | |
| ε Lyr Double Double | 18 | 44.3 | +39 | 40 | | Multiple Star | 142 |

| Designation/Name | RA h | m | Dec ° | ' | Mag. | Type | Page |
|---|---|---|---|---|---|---|---|
| ζ Lyr | 18 | 44.8 | +37 | 36 | | Double Star | 138 |
| M57 NGC 6720 Ring Nebula | 18 | 53.6 | +33 | 02 | 8.8 | Planetary Nebula | 150 |
| M56 NGC 6779 | 19 | 16.6 | +30 | 11 | 8.3 | Globular Cluster | 89 |
| **Monoceros (Mon)** | | | | | | | |
| β Mon | 06 | 28.8 | −07 | 02 | | Triple Star | 142 |
| NGC 2237−2239, NGC 2246 Rosette Nebula | 06 | 32.3 | +05 | 03 | 10 | Emission Nebula | 104 |
| NGC 2244 | 06 | 32.4 | +04 | 52 | 4.8 | Open Cluster | 123 |
| NGC 2261 Hubble's Variable Nebula | 06 | 39.2 | +08 | 44 | c.10 | Reflection Nebula | 107 |
| M50 NGC 2323 | 07 | 03.2 | −08 | 20 | 5.9 | Open Cluster | 122 |
| **Musca (Mus)** | | | | | | | |
| NGC 4833 | 12 | 59.6 | −70 | 53 | 6.9 | Globular Cluster | 87 |
| **Ophiuchus (Oph)** | | | | | | | |
| M107 NGC 6171 | 16 | 32.5 | −13 | 03 | 8.1 | Globular Cluster | 87 |
| M12 NGC 6218 | 16 | 47.2 | −01 | 57 | 6.8 | Globular Cluster | 85 |
| M10 NGC 6254 | 16 | 57.1 | −04 | 06 | 6.6 | Globular Cluster | 85 |
| M62 NGC 6266 | 17 | 01.2 | −30 | 07 | 6.7 | Globular Cluster | 86 |
| M19 NGC 6273 | 17 | 02.6 | −26 | 16 | 6.7 | Globular Cluster | 85 |
| B64 | 17 | 17.2 | −18 | 33 | | Dark Nebula | 110 |
| M9 NGC 6333 | 17 | 19.2 | −18 | 31 | 7.6 | Globular Cluster | 86 |
| B59 | 17 | 21.0 | −27 | 00 | | Dark Nebula | 108 |
| B72 | 17 | 23.5 | −23 | 28 | | Dark Nebula | 110 |
| B78 | 17 | 33.0 | −26 | 00 | | Dark Nebula | 108 |
| M14 NGC 6402 | 17 | 37.6 | −03 | 15 | 7.6 | Globular Cluster | 87 |
| IC 4665 | 17 | 46 | +05 | 43 | 4.2 | Open Cluster | 126 |
| NGC 6572 | 18 | 12.1 | +06 | 51 | 8.1 | Planetary Nebula | 155 |
| **Orion (Ori)** | | | | | | | |
| θ¹ Ori Trapezium | 05 | 35.3 | −05 | 23 | | Multiple Star | 141 |
| M42 NGC 1976 Orion Nebula | 05 | 35.4 | −05 | 27 | 4.0 | Emission Nebula | 99 |
| M43 NGC 1982 Ori | 05 | 35.6 | −05 | 16 | 5.0 | Emission Nebula | 99 |
| ζ Ori (Alnitak) | 05 | 40.8 | −01 | 57 | | Double Star | 143 |
| M78 NGC 2067/206 | 05 | 46.7 | +00 | 03 | 8 | Reflection Nebula | 105 |
| **Pavo (Pav)** | | | | | | | |
| NGC 6752 | 19 | 10.9 | −59 | 59 | 5.4 | Globular Cluster | 78 |
| **Pegasus (Peg)** | | | | | | | |
| M15 NGC 7078 Peg | 21 | 30.0 | +12 | 10 | 6.0 | Globular Cluster | 83 |
| **Perseus (Per)** | | | | | | | |
| M76 NGC 650/651 Little Dumbbell | 01 | 42.4 | +51 | 34 | 10.1 | Planetary Nebula | 153 |
| NGC 869 } Double Cluster { | 02 | 19.0 | +57 | 09 | 5.3 | } Open Clusters { | 118 |
| NGC 884 } | 02 | 22.4 | +57 | 07 | 6.1 | | 118 |
| M34 NGC 1039 | 02 | 42.0 | +42 | 47 | 5.2 | Open Cluster | 120 |
| NGC 1499 California Nebula | 04 | 00.7 | +36 | 37 | 9 | Emission Nebula | 103 |
| NGC 1528 | 04 | 15.4 | +51 | 14 | 6.4 | Open Cluster | 120 |
| **Pisces (Psc)** | | | | | | | |
| M74 NGC 628 | 01 | 36.7 | +15 | 47 | 9.4 | Galaxy | 69 |
| α Psc | 02 | 02.0 | +02 | 46 | | Double Star | 145 |

| Designation/Name | RA h | m | Dec ° | ′ | Mag. | Type | Page |
|---|---|---|---|---|---|---|---|
| **Puppis (Pup)** | | | | | | | |
| M47 NGC 2422 | 07 | 36.6 | −14 | 30 | 4.4 | Open Cluster | 124 |
| NGC 2438 | 07 | 41.8 | −14 | 44 | 11. | Planetary Nebula | 158 |
| M46 NGC 2437 | 07 | 41.8 | −14 | 49 | 6.1 | Open Cluster | 124 |
| NGC 2440 | 07 | 41.9 | −18 | 13 | 9.4 | Planetary Nebula | 158 |
| M93 NGC 2447 | 07 | 44.6 | −23 | 52 | 6.2 | Open Cluster | 125 |
| **Sagitta (Sge)** | | | | | | | |
| M71 NGC 6838 | 19 | 53.8 | +18 | 47 | 8.3 | Globular Cluster | 90 |
| **Sagittarius (Sgr)** | | | | | | | |
| M20 NGC 6514 Trifid Nebula | 18 | 02.3 | −23 | 02 | 6.3 | Emission Nebula | 101 |
| M8 NGC 6523 Lagoon Nebula | 18 | 03.8 | −24 | 21 | 6.0 | Emission Nebula | 100 |
| M24 Small Sagittarius Star Cloud | 18 | 16.5 | −18 | 50 | 4.6 | Open Cluster | 127 |
| B92 | 18 | 16.9 | −18 | 02 | | Dark Nebula | 110 |
| M17 NGC 6618 Swan Nebula | 18 | 20.8 | −16 | 11 | 6.0 | Emission Nebula | 102 |
| M28 NGC 6626 | 18 | 24.5 | −24 | 52 | 6.8 | Globular Cluster | 78 |
| M69 NGC 6637 | 18 | 31.4 | −32 | 21 | 7.6 | Globular Cluster | 89 |
| M25 IC 4725 | 18 | 31.6 | −19 | 15 | 4.6 | Open Cluster | 127 |
| M22 NGC 6656 | 18 | 36.4 | −23 | 54 | 5.1 | Globular Cluster | 77 |
| M70 NGC 6681 | 18 | 42.2 | −32 | 18 | 8.0 | Globular Cluster | 89 |
| M54 NGC 6715 | 18 | 55.1 | −30 | 29 | 7.6 | Globular Cluster | 90 |
| M55 NGC 6809 | 19 | 40.0 | −30 | 58 | 6.4 | Globular Cluster | 84 |
| M75 NGC 6864 | 20 | 06.1 | −21 | 55 | 8.5 | Globular Cluster | 84 |
| **Scorpius (Sco)** | | | | | | | |
| β Sco | 16 | 05.4 | −19 | 48 | | Double Star | 139 |
| ν Sco | 16 | 12.0 | −19 | 28 | | Multiple Star | 143 |
| M80 NGC 6093 | 16 | 17.0 | −22 | 59 | 7.3 | Globular Cluster | 81 |
| M4 NGC 6121 | 16 | 23.6 | −26 | 32 | 5.8 | Globular Cluster | 81 |
| M6 NGC 6405 Butterfly Cluster | 17 | 40.1 | −32 | 13 | 4.2 | Open Cluster | 126 |
| M7 NGC 6475 | 17 | 53.9 | −34 | 49 | 3.3 | Open Cluster | 127 |
| α Sco (Antares) | 16 | 29.4 | −26 | 26 | | Double Star | 145 |
| **Sculptor (Scl)** | | | | | | | |
| NGC 253 Silver Coin | 00 | 47.6 | −25 | 17 | 7.8 | Galaxy | 70 |
| **Scutum (Sct)** | | | | | | | |
| M11 NGC 6705 Wild Duck | 18 | 51.1 | −06 | 16 | 5.8 | Open Cluster | 128 |
| NGC 6712 | 18 | 53.1 | −08 | 42 | 8.2 | Globular Cluster | 91 |
| **Serpens (Ser)** | | | | | | | |
| M5 NGC 5904 | 15 | 18.6 | +02 | 05 | 5.7 | Globular Cluster | 80 |
| M16 IC 4703/NGC 6611 Eagle Nebula | 18 | 18.8 | −13 | 47 | 6.0 | Emission Nebula | 102 |
| θ Ser | 18 | 56.2 | +04 | 12 | | Double Star | 138 |
| **Sextans (Sex)** | | | | | | | |
| NGC 3115 Spindle Galaxy | 10 | 05.2 | −07 | 43 | 8.9 | Galaxy | 70 |
| **Taurus (Tau)** | | | | | | | |
| NGC 1435 Merope Nebula | 03 | 46.1 | +25 | 47 | | Reflection Nebula | 107 |
| M45 Pleiades | 03 | 47.0 | +24 | 07 | 1.2 | Open Cluster | 114 |
| Hyades | 04 | 27 | +16 | | 0.5 | Open Cluster | 116 |
| M1 NGC 1952 Crab Nebula | 05 | 34.5 | +22 | 01 | 8.4 | Supernova Remnant | 163 |

| Designation/Name | RA h | m | Dec ° | , | Mag. | Type | Page |
|---|---|---|---|---|---|---|---|
| **Triangulum (Tri)** | | | | | | | |
| M33 NGC 598 | 01 | 33.9 | +30 | 39 | 5.7 | Galaxy | 55 |
| **Tucana (Tuc)** | | | | | | | |
| 47 Tucanae NGC 104 | 00 | 24.1 | −72 | 05 | 4.0 | Globular Cluster | 77 |
| Small Magellanic Cloud | 00 | 52.7 | −72 | 50 | 2.3 | Galaxy | 51 |
| **Ursa Major (UMa)** | | | | | | | |
| M81 NGC 3031 | 09 | 55.6 | +69 | 04 | 6.9 | Galaxy | 56 |
| M82 NGC 3034 | 09 | 55.8 | +69 | 41 | 8.4 | Galaxy | 56 |
| M97 NGC 3587 The Owl Nebula | 11 | 14.8 | +55 | 01 | 9.9 | Planetary Nebula | 152 |
| ξ UMa | 11 | 18.2 | +31 | 32 | | Double Star | 143 |
| M40 | 12 | 22.4 | +58 | 05 | | Asterism | 131 |
| ζ UMa (Mizar) | 13 | 23.9 | +54 | 56 | | Double Star | 136 |
| M101 NGC 5457 | 14 | 03.2 | +54 | 21 | 7.9 | Galaxy | 61 |
| **Vela (Vel)** | | | | | | | |
| IC 2391 | 08 | 40.2 | −53 | 04 | 2.5 | Open Cluster | 126 |
| NGC 2736 | 09 | 00.4 | −45 | 54 | | Supernova Remnant | 167 |
| **Virgo (Vir)** | | | | | | | |
| M61 NGC 4303 | 12 | 21.9 | +04 | 28 | 9.7 | Galaxy | 68 |
| M84 NGC 4374 | 12 | 25.1 | +12 | 53 | 9.1 | Galaxy | 66 |
| M86 NGC 4406 | 12 | 26.2 | +12 | 57 | 8.9 | Galaxy | 66 |
| M49 NGC 4472 | 12 | 29.8 | +08 | 00 | 8.4 | Galaxy | 68 |
| M87 NGC 4486 | 12 | 30.8 | +12 | 24 | 8.6 | Galaxy | 66 |
| M89 NGC 4552 | 12 | 35.7 | +12 | 33 | 9.8 | Galaxy | 63 |
| M90 NGC 4569 | 12 | 36.8 | +13 | 10 | 9.5 | Galaxy | 63 |
| M58 NGC 4579 | 12 | 37.7 | +11 | 49 | 9.7 | Galaxy | 63 |
| M104 NGC 4594 Sombrero Galaxy | 12 | 40.0 | −11 | 37 | 8.0 | Galaxy | 67 |
| M59 NGC 4621 | 12 | 42.0 | +11 | 39 | 9.6 | Galaxy | 63 |
| M60 NGC 4649 | 12 | 43.7 | +11 | 33 | 8.8 | Galaxy | 63 |
| **Vulpecula (Vul)** | | | | | | | |
| Cr 399 Coathanger | 19 | 25.4 | +20 | 11 | | Asterism | 131 |
| M27 NGC 6583 Dumbbell Nebula | 19 | 59.6 | +22 | 43 | 7.3 | Planetary Nebula | 149 |

## The Messier List

Probably the most widely used deep sky catalog for amateur observers, Messier's list features many of the brightest objects in the sky. Many observers consider it an interesting challenge to "collect" each of the Messier objects.

| The Messier List | | | | | | | |
|---|---|---|---|---|---|---|---|
| Designation/Name | Constellation | RA h | m | Dec ° | ' | Mag. | Type | Page |
| M1 NGC 1952 Crab Nebula | Tau | 05 | 34.5 | +22 | 01 | 8.4 | Supernova Remnant | 163 |
| M2 NGC 7089 | Aqr | 21 | 33.5 | –00 | 49 | 6.4 | Globular Cluster | 83 |
| M3 NGC 5272 | CVn | 13 | 42.2 | +28 | 23 | 5.9 | Globular Cluster | 81 |
| M4 NGC 6121 | Sco | 16 | 23.6 | –26 | 32 | 5.8 | Globular Cluster | 81 |
| M5 NGC 5904 | Ser | 15 | 18.6 | +02 | 05 | 5.7 | Globular Cluster | 80 |
| M6 NGC 6405 Butterfly Cluster | Sco | 17 | 40.1 | –32 | 13 | 4.2 | Open Cluster | 126 |
| M7 NGC 6475 | Sco | 17 | 53.9 | –34 | 49 | 3.3 | Open Cluster | 127 |
| M8 NGC 6523 Lagoon Nebula | Sgr | 18 | 03.8 | –24 | 21 | 6.0 | Emission Nebula | 100 |
| M9 NGC 6333 | Oph | 17 | 19.2 | –18 | 31 | 7.6 | Globular Cluster | 86 |
| M10 NGC 6254 | Oph | 16 | 57.1 | –04 | 06 | 6.6 | Globular Cluster | 85 |
| M11 NGC 6705 Wild Duck | Sct | 18 | 51.1 | –06 | 16 | 5.8 | Open Cluster | 128 |
| M12 NGC 6218 | Oph | 16 | 47.2 | –01 | 57 | 6.8 | Globular Cluster | 85 |
| M13 NGC 6205 | Her | 16 | 41.7 | +36 | 28 | 5.7 | Globular Cluster | 79 |
| M14 NGC 6402 | Oph | 17 | 37.6 | –03 | 15 | 7.6 | Globular Cluster | 87 |
| M15 NGC 7078 | Peg | 21 | 30.0 | +12 | 10 | 6.0 | Globular Cluster | 83 |
| M16 IC 4703/ NGC 6611 Eagle Nebula | Ser | 18 | 18.8 | –13 | 47 | 6.0 | Emission Nebula | 102 |
| M17 NGC 6618 Swan Nebula | Sgr | 18 | 20.8 | –16 | 11 | 6.0 | Emission Nebula | 102 |
| M18 NGC 6613 | Sgr | 18 | 19.9 | –17 | 08 | 6.9 | Open Cluster | |
| M19 NGC 6273 | Oph | 17 | 02.6 | –26 | 16 | 6.7 | Globular Cluster | 85 |
| M20 NGC 6514 Trifid Nebula | Sgr | 18 | 02.3 | –23 | 02 | 6.3 | Emission Nebula | 101 |
| M21 NGC 6531 | Sgr | 18 | 04.6 | –22 | 30 | 6.5 | Open cluster | |
| M22 NGC 6656 | Sgr | 18 | 36.4 | –23 | 54 | 5.1 | Globular Cluster | 77 |
| M23 NGC 6494 | Sgr | 17 | 56.8 | –19 | 01 | 6.9 | Open Cluster | |
| M24 Small Sagittarius Star Cloud | Sgr | 18 | 16.5 | –18 | 50 | 4.6 | Open Cluster | 127 |
| M25 IC 4725 | Sgr | 18 | 31.6 | –19 | 15 | 4.6 | Open Cluster | 127 |
| M26 NGC 6694 | Sct | 18 | 45.2 | –09 | 24 | 8.0 | Open Cluster | |
| M27 NGC 6583 Dumbbell Nebula | Vul | 19 | 59.6 | +22 | 43 | 7.3 | Planetary Nebula | 149 |
| M28 NGC 6626 | Sgr | 18 | 24.5 | –24 | 52 | 6.8 | Globular Cluster | 78 |
| M29 NGC 6913 | Cyg | 20 | 23.9 | +38 | 32 | 6.6 | Open Cluster | 130 |
| M30 NGC 7099 | Cap | 21 | 40.4 | –23 | 11 | 7.3 | Globular Cluster | 84 |

| Designation/Name | Constellation | RA h | m | Dec ° | ′ | Mag. | Type | Page |
|---|---|---|---|---|---|---|---|---|
| M31 NGC 224 Andromeda Galaxy | And | 00 | 42.7 | +41 | 16 | 3.4 | Galaxy | 53 |
| M32 NGC 221 | And | 00 | 42.7 | +40 | 52 | 8.1 | Galaxy | 53 |
| M33 NGC 598 | Tri | 01 | 33.9 | +30 | 39 | 5.7 | Galaxy | 55 |
| M34 NGC 1039 | Per | 02 | 42.0 | +42 | 47 | 5.2 | Open Cluster | 120 |
| M35 NGC 2168 | Gem | 06 | 08.9 | +24 | 20 | 5.1 | Open Cluster | 122 |
| M36 NGC 1960 | Aur | 05 | 36.1 | +34 | 08 | 6.0 | Open Cluster | 120 |
| M37 NGC 2099 | Aur | 05 | 52.4 | +32 | 33 | 5.6 | Open Cluster | 121 |
| M38 NGC 1912 | Aur | 05 | 28.7 | +39 | 50 | 6.4 | Open Cluster | 120 |
| M39 NGC 7092 | Cyg | 21 | 32.2 | +48 | 26 | 5.0 | Open Cluster | 130 |
| M40 | UMa | 12 | 22.4 | +58 | 05 | | Asterism | 131 |
| M41 NGC 2287 | CMa | 06 | 47.0 | −20 | 44 | 4.5 | Open Cluster | 123 |
| M42 NGC 1976 Orion Nebula | Ori | 05 | 35.4 | −05 | 27 | 4.0 | Emission Nebula | 99 |
| M43 NGC 1982 | Ori | 05 | 35.6 | −05 | 16 | 5.0 | Emission Nebula | 99 |
| M44 NGC 2632 Praesepe | Cnc | 08 | 40.1 | +19 | 59 | 3.1 | Open Cluster | 117 |
| M45 Pleiades | Tau | 03 | 47.0 | +24 | 07 | 1.2 | Open Cluster | 114 |
| M46 NGC 2437 | Pup | 07 | 41.8 | −14 | 49 | 6.1 | Open Cluster | 124 |
| M47 NGC 2422 | Pup | 07 | 36.6 | −14 | 30 | 4.4 | Open Cluster | 124 |
| M48 NGC 2458 | Hya | 08 | 13.8 | −05 | 48 | 5.8 | Open Cluster | 125 |
| M49 NGC 4472 | Vir | 12 | 29.8 | +08 | 00 | 8.4 | Galaxy | 68 |
| M50 NGC 2323 | Mon | 07 | 03.2 | −08 | 20 | 5.9 | Open Cluster | 122 |
| M51 NGC 5194 Whirlpool Galaxy | CVn | 13 | 29.9 | +47 | 12 | 8.4 | Galaxy | 60 |
| M52 NGC 7654 | Cas | 23 | 24.2 | +61 | 35 | 6.3 | Open Cluster | 128 |
| M53 NGC 5024 | Com | 13 | 12.9 | +18 | 10 | 7.5 | Globular Cluster | 82 |
| M54 NGC 6715 | Sgr | 18 | 55.1 | −30 | 29 | 7.6 | Globular Cluster | 90 |
| M55 NGC 6809 | Sgr | 19 | 40.0 | −30 | 58 | 6.4 | Globular Cluster | 84 |
| M56 NGC 6779 | Lyr | 19 | 16.6 | +30 | 11 | 8.3 | Globular Cluster | 89 |
| M57 NGC 6720 Ring Nebula | Lyr | 18 | 53.6 | +33 | 02 | 8.8 | Planetary Nebula | 150 |
| M58 NGC 4579 | Vir | 12 | 37.7 | +11 | 49 | 9.7 | Galaxy | 63 |
| M59 NGC 4621 | Vir | 12 | 42.0 | +11 | 39 | 9.6 | Galaxy | 63 |
| M60 NGC 4649 | Vir | 12 | 43.7 | +11 | 33 | 8.8 | Galaxy | 63 |
| M61 NGC 4303 | Vir | 12 | 21.9 | +04 | 28 | 9.7 | Galaxy | 68 |
| M62 NGC 6266 | Oph | 17 | 01.2 | −30 | 07 | 6.7 | Globular Cluster | 86 |
| M63 NGC 5055 Sunflower Galaxy | CVn | 13 | 15.8 | +42 | 02 | 8.6 | Galaxy | 61 |
| M64 NGC 4826 Black Eye Galaxy | Com | 12 | 56.7 | +21 | 41 | 8.5 | Galaxy | 62 |
| M65 NGC 3623 | Leo | 11 | 18.9 | +13 | 05 | 9.3 | Galaxy | 58 |
| M66 NGC 3627 | Leo | 11 | 20.2 | +12 | 59 | 8.9 | Galaxy | 58 |
| M67 NGC 2682 | Cnc | 08 | 50.4 | +11 | 49 | 6.9 | Cluster | 118 |
| M68 NGC 4590 | Hya | 12 | 39.5 | −26 | 45 | 7.7 | Globular Cluster | 89 |
| M69 NGC 6637 | Sgr | 18 | 31.4 | −32 | 21 | 7.6 | Globular Cluster | 89 |
| M70 NGC 6681 | Sgr | 18 | 42.2 | −32 | 18 | 8.0 | Globular Cluster | 89 |
| M71 NGC 6838 | Sge | 19 | 53.8 | +18 | 47 | 8.3 | Globular Cluster | 90 |

| Designation/Name | Constellation | RA h | m | Dec ° | ′ | Mag. | Type | Page |
|---|---|---|---|---|---|---|---|---|
| M72 NGC 6981 | Aqr | 20 | 53.5 | −12 | 32 | 9.3 | Globular Cluster | 83 |
| M73 NGC 6994 | Aqr | 20 | 59.0 | −12 | 58 | | Asterism | 131 |
| M74 NGC 628 | Psc | 01 | 36.7 | +15 | 47 | 9.4 | Galaxy | 69 |
| M75 NGC 6864 | Sgr | 20 | 06.1 | −21 | 55 | 8.5 | Globular Cluster | 84 |
| M76 NGC 650/651 Little Dumbbell | Per | 01 | 42.4 | +51 | 34 | 10.1 | Planetary Nebula | 153 |
| M77 NGC 1068 | Cet | 02 | 42.7 | −00 | 01 | 8.9 | Galaxy | 69 |
| M78 NGC 2067/2068 | Ori | 05 | 46.7 | +00 | 03 | 8 | Reflection Nebula | 105 |
| M79 NGC 1904 | Lep | 05 | 24.5 | −24 | 33 | 7.8 | Globular Cluster | 87 |
| M80 NGC 6093 | Sco | 16 | 17.0 | −22 | 59 | 7.3 | Globular Cluster | 81 |
| M81 NGC 3031 | UMa | 09 | 55.6 | +69 | 04 | 6.9 | Galaxy | 56 |
| M82 NGC 3034 | UMa | 09 | 55.8 | +69 | 41 | 8.4 | Galaxy | 56 |
| M83 NGC 5236 | Hya | 13 | 37.0 | −29 | 52 | 7.6 | Galaxy | 70 |
| M84 NGC 4374 | Vir | 12 | 25.1 | +12 | 53 | 9.1 | Galaxy | 66 |
| M85 NGC 4382 | Com | 12 | 25.4 | +18 | 11 | 9.1 | Galaxy | 64 |
| M86 NGC 4406 | Vir | 12 | 26.2 | +12 | 57 | 8.9 | Galaxy | 66 |
| M87 NGC 4486 | Vir | 12 | 30.8 | +12 | 24 | 8.6 | Galaxy | 66 |
| M88 NGC 4501 | Com | 12 | 32.0 | +14 | 25 | 9.6 | Galaxy | 64 |
| M89 NGC 4552 | Vir | 12 | 35.7 | +12 | 33 | 9.8 | Galaxy | 63 |
| M90 NGC 4569 | Vir | 12 | 36.8 | +13 | 10 | 9.5 | Galaxy | 63 |
| M91 NGC 4548 | Com | 12 | 35.4 | +14 | 30 | 10.2 | Galaxy | 64 |
| M92 NGC 6341 | Her | 17 | 17.1 | +43 | 08 | 6.5 | Globular Cluster | 80 |
| M93 NGC 2447 | Pup | 07 | 44.6 | −23 | 52 | 6.2 | Open Cluster | 125 |
| M94 NGC 4736 | CVn | 12 | 50.9 | +41 | 07 | 8.2 | Galaxy | 62 |
| M95 NGC 3351 | Leo | 10 | 44.0 | +11 | 42 | 9.7 | Galaxy | 59 |
| M96 NGC 3368 | Leo | 10 | 46.8 | +10 | 49 | 9.2 | Galaxy | 59 |
| M97 NGC 3587 Owl Nebula | UMa | 11 | 14.8 | +55 | 01 | 9.9 | Planetary Nebula | 152 |
| M98 NGC 4192 | Com | 12 | 13.8 | +14 | 54 | 10.1 | Galaxy | 64 |
| M99 NGC 4254 | Com | 12 | 18.8 | +14 | 25 | 9.9 | Galaxy | 64 |
| M100 NGC 4321 | Com | 12 | 22.6 | +15 | 47 | 9.3 | Galaxy | 64 |
| M101 NGC 5457 | UMa | 14 | 03.2 | +54 | 21 | 7.9 | Galaxy | 61 |
| M102 Duplicate of M101 | | | | | | | | |
| M103 NGC 581 | Cas | 01 | 33.2 | +60 | 52 | 7.4 | Open Cluster | 129 |
| M104 NGC 4594 Sombrero Galaxy | Vir | 12 | 40.0 | −11 | 37 | 8.0 | Galaxy | 67 |
| M105 NGC 3379 | Leo | 10 | 47.8 | +12 | 35 | 9.3 | Galaxy | 59 |
| M106 NGC 4258 | CVn | 12 | 19.0 | +47 | 18 | 8.4 | Galaxy | |
| M107 NGC 6171 | Oph | 16 | 32.5 | −13 | 03 | 8.1 | Globular Cluster | 87 |
| M108 NGC 3556 | UMa | 11 | 11.5 | +55 | 40 | 10.0 | Galaxy | |
| M109 NGC 3992 | UMa | 11 | 57.6 | +53 | 23 | 9.8 | Galaxy | |
| M110 NGC 205 | And | 00 | 40.4 | +41 | 41 | 8.1 | Galaxy | 53 |

## *Objects by Magnitude*

The objects described in this book range from those easily visible to the naked eye, to faint nebulae, distant globular clusters and galaxies that can only be seen with larger instruments. A challenge for the observer is to see which is the faintest object that can be detected.

| | | | | | | | |
|---|---|---|---|---|---|---|---|
| **Objects by Magnitude** | | | | | | | |
| Designation/Name | Constellation | RA h | m | Dec °     ' | Mag. | Type | Page |
| Large Magellanic Cloud | Dor | 05 | 23.6 | –69 45 | 0.1 | Galaxy | 51 |
| Hyades | Tau | 04 | 27 | +16 | 0.5 | Open Cluster | 116 |
| NGC 3372 Eta Carinae Nebula | Car | 10 | 43.8 | –59 52 | 1.0 | Emission Nebula | 98 |
| M45 Pleiades | Tau | 03 | 47.0 | +24 07 | 1.2 | Open Cluster | 114 |
| Mel 111 | Com | 12 | 25 | +26 | 1.8 | Open Cluster | 116 |
| IC 2602 Southern Pleiades | Car | 10 | 43.2 | –64 24 | 1.9 | Open Cluster | 117 |
| Small Magellanic Cloud | Tuc | 00 | 52.7 | –72 50 | 2.3 | Galaxy | 51 |
| IC 2391 | Vel | 08 | 40.2 | –53 04 | 2.5 | Open Cluster | 126 |
| M44 NGC 2632 Praesepe | Cnc | 08 | 40.1 | +19 59 | 3.1 | Open Cluster | 117 |
| M7 NGC 6475 | Sco | 17 | 53.9 | –34 49 | 3.3 | Open Cluster | 127 |
| M31 NGC 224 Andromeda Galaxy | And | 00 | 42.7 | +41 16 | 3.4 | Galaxy | 53 |
| NGC 5139 Omega Centauri | Cen | 13 | 26.8 | –47 29 | 3.5 | Globular Cluster | 76 |
| 47 Tucanae NGC 104 | Tuc | 00 | 24.1 | –72 05 | 4.0 | Globular Cluster | 77 |
| M42 NGC 1976 Orion Nebula | Ori | 05 | 35.4 | –05 27 | 4.0 | Emission Nebula | 99 |
| NGC 2362 Tau CMa Cluster | CMa | 07 | 18.8 | –26 47 | 4.1 | Open Cluster | 124 |
| NGC 4755 Jewel Box | Cru | 15 | 53.6 | –60 20 | 4.2 | Open Cluster | 126 |
| M6 NGC 6405 Butterfly Cluster | Sco | 17 | 40.1 | –32 13 | 4.2 | Open Cluster | 126 |
| IC 4665 | Oph | 17 | 46 | +05 43 | 4.2 | Open Cluster | 126 |
| M47 NGC 2422 | Pup | 07 | 36.6 | –14 30 | 4.4 | Open Cluster | 124 |
| M41 NGC 2287 | CMa | 06 | 47.0 | –20 44 | 4.5 | Open Cluster | 123 |
| M24 Small Sagittarius Star Cloud | Sgr | 18 | 16.5 | –18 50 | 4.6 | Open Cluster | 127 |
| M25 IC 4725 | Sgr | 18 | 31.6 | –19 15 | 4.6 | Open Cluster | 127 |
| NGC 2244 | Mon | 06 | 32.4 | +04 52 | 4.8 | Open Cluster | 123 |
| M43 NGC 1982 Ori | Ori | 05 | 35.6 | –05 16 | 5.0 | Emission Nebula | 99 |
| NGC 7000 North America Nebula | Cyg | 20 | 58.8 | +44 20 | 5? | Emission Nebula | 104 |
| M39 NGC 7092 | Cyg | 21 | 32.2 | +48 26 | 5.0 | Open Cluster | 130 |
| M35 NGC 2168 | Gem | 06 | 08.9 | +24 20 | 5.1 | Open Cluster | 122 |
| M22 NGC 6656 | Sgr | 18 | 36.4 | –23 54 | 5.1 | Globular Cluster | 77 |
| M34 NGC 1039 | Per | 02 | 42.0 | +42 47 | 5.2 | Open Cluster | 120 |
| NGC 869 | Per | 02 | 19.0 | +57 09 | 5.3 | Open Cluster | 118 |

| Designation/Name | Constellation | RA h | m | Dec ° | , | Mag. | Type | Page |
|---|---|---|---|---|---|---|---|---|
| NGC 6752 | Pav | 19 | 10.9 | −59 | 59 | 5.4 | Globular Cluster | 78 |
| M37 NGC 2099 | Aur | 05 | 52.4 | +32 | 33 | 5.6 | Open Cluster | 121 |
| M33 NGC 598 | Tri | 01 | 33.9 | +30 | 39 | 5.7 | Galaxy | 55 |
| NGC 752 | And | 01 | 57.8 | 37 | 41 | 5.7 | Open Cluster | 130 |
| M13 NGC 6205 | Her | 16 | 41.7 | +36 | 28 | 5.7 | Globular Cluster | 79 |
| M5 NGC 5904 | Ser | 15 | 18.6 | +02 | 05 | 5.7 | Globular Cluster | 80 |
| M48 NGC 2458 | Hya | 08 | 13.8 | −05 | 48 | 5.8 | Open Cluster | 125 |
| M4 NGC 6121 | Sco | 16 | 23.6 | −26 | 32 | 5.8 | Globular Cluster | 81 |
| M11 NGC 6705 Wild Duck | Sct | 18 | 51.1 | −06 | 16 | 5.8 | Open Cluster | 128 |
| M50 NGC 2323 | Mon | 07 | 03.2 | −08 | 20 | 5.9 | Open Cluster | 122 |
| M3 NGC 5272 | CVn | 13 | 42.2 | +28 | 23 | 5.9 | Globular Cluster | 81 |
| NGC 6397 | Ara | 17 | 40.7 | −53 | 40 | 5.9 | Globular Cluster | 81 |
| M36 NGC 1960 | Aur | 05 | 36.1 | +34 | 08 | 6.0 | Open Cluster | 120 |
| M8 NGC 6523 Lagoon Nebula | Sgr | 18 | 03.8 | −24 | 21 | 6.0 | Emission Nebula | 100 |
| M16 IC 4703/NGC 6611 Eagle Nebula | Ser | 18 | 18.8 | −13 | 47 | 6.0 | Emission Nebula | 102 |
| M17 NGC 6618 Swan Nebula | Sgr | 18 | 20.8 | −16 | 11 | 6.0 | Emission Nebula | 102 |
| M15 NGC 7078 Peg | Peg | 21 | 30.0 | +12 | 10 | 6.0 | Globular Cluster | 83 |
| NGC 884 | Per | 02 | 22.4 | +57 | 07 | 6.1 | Open Cluster | 118 |
| M46 NGC 2437 | Pup | 07 | 41.8 | −14 | 49 | 6.1 | Open Cluster | 124 |
| M93 NGC 2447 | Pup | 07 | 44.6 | −23 | 52 | 6.2 | Open Cluster | 125 |
| M20 NGC 6514 Trifid Nebula | Sgr | 18 | 02.3 | −23 | 02 | 6.3 | Emission Nebula | 101 |
| M52 NGC 7654 | Cas | 23 | 24.2 | +61 | 35 | 6.3 | Open Cluster | 128 |
| NGC 457 | Cas | 01 | 19.1 | +58 | 20 | 6.4 | Open Cluster | 129 |
| M38 NGC 1912 | Aur | 05 | 28.7 | +39 | 50 | 6.4 | Open Cluster | 120 |
| NGC 1528 | Per | 04 | 15.4 | +51 | 14 | 6.4 | Open Cluster | 120 |
| M55 NGC 6809 | Sgr | 19 | 40.0 | −30 | 58 | 6.4 | Globular Cluster | 84 |
| M2 NGC 7089 | Aqr | 21 | 33.5 | −00 | 49 | 6.4 | Globular Cluster | 83 |
| M92 NGC 6341 | Her | 17 | 17.1 | +43 | 08 | 6.5 | Globular Cluster | 80 |
| M10 NGC 6254 | Oph | 16 | 57.1 | −04 | 06 | 6.6 | Globular Cluster | 85 |
| M29 NGC 6913 | Cyg | 20 | 23.9 | +38 | 32 | 6.6 | Open Cluster | 130 |
| NGC 5128 | Cen | 13 | 25.5 | −43 | 01 | 6.7 | Galaxy | 58 |
| M62 NGC 6266 | Oph | 17 | 01.2 | −30 | 07 | 6.7 | Globular Cluster | 86 |
| M19 NGC 6273 | Oph | 17 | 02.6 | −26 | 16 | 6.7 | Globular Cluster | 85 |
| M12 NGC 6218 | Oph | 16 | 47.2 | −01 | 57 | 6.8 | Globular Cluster | 85 |
| M28 NGC 6626 | Sgr | 18 | 24.5 | −24 | 52 | 6.8 | Globular Cluster | 78 |
| M67 NGC 2682 | Cnc | 08 | 50.4 | +11 | 49 | 6.9 | Open Cluster | 118 |
| M81 NGC 3031 | UMa | 09 | 55.6 | +69 | 04 | 6.9 | Galaxy | 56 |
| NGC 4833 | Mus | 12 | 59.6 | 70 | 53 | 6.9 | Globular Cluster | 87 |
| M80 NGC 6093 | Sco | 16 | 17.0 | −22 | 59 | 7.3 | Globular Cluster | 81 |
| M27 NGC 6583 Dumbbell Nebula | Vul | 19 | 59.6 | +22 | 43 | 7.3 | Planetary Nebula | 149 |
| M30 NGC 7099 | Cap | 21 | 40.4 | −23 | 11 | 7.3 | Globular Cluster | 84 |

| Designation/Name | Constellation | RA h | m | Dec ° | ′ | Mag. | Type | Page |
|---|---|---|---|---|---|---|---|---|
| NGC 7293 Helix Nebula | Aqr | 22 | 29.6 | –20 | 48 | 7.3 | Planetary Nebula | 157 |
| M103 NGC 581 | Cas | 01 | 33.2 | +60 | 52 | 7.4 | Open Cluster | 129 |
| M53 NGC 5024 | Com | 13 | 12.9 | +18 | 10 | 7.5 | Globular Cluster | 82 |
| M14 NGC 6402 | Oph | 17 | 37.6 | –03 | 15 | 7.6 | Globular Cluster | 87 |
| M69 NGC 6637 | Sgr | 18 | 31.4 | –32 | 21 | 7.6 | Globular Cluster | 89 |
| M54 NGC 6715 | Sgr | 18 | 55.1 | –30 | 29 | 7.6 | Globular Cluster | 90 |
| M83 NGC 5236 | Hya | 13 | 37.0 | –29 | 52 | 7.6 | Galaxy | 70 |
| M9 NGC 6333 | Oph | 17 | 19.2 | –18 | 31 | 7.6 | Globular Cluster | 86 |
| M68 NGC 4590 | Hya | 12 | 39.5 | –26 | 45 | 7.7 | Globular Cluster | 89 |
| NGC 253 Silver Coin | Scl | 00 | 47.6 | –25 | 17 | 7.8 | Galaxy | 70 |
| M79 NGC 1904 | Lep | 05 | 24.5 | –24 | 33 | 7.8 | Globular Cluster | 87 |
| NGC 3242 Ghost of Jupiter | Hya | 10 | 24.8 | –18 | 38 | 7.8 | Planetary Nebula | 154 |
| M101 NGC 5457 | UMa | 14 | 03.2 | +54 | 21 | 7.9 | Galaxy | 61 |
| M78 NGC 2067/206 | Ori | 05 | 46.7 | +00 | 03 | 8 | Reflection Nebula | 105 |
| NGC 2070 Tarantula Nebula | Dor | 05 | 38.7 | –69 | 06 | 8.0(?) | Emission Nebula | 103 |
| M104 NGC 4594 Sombrero Galaxy | Vir | 12 | 40.0 | –11 | 37 | 8.0 | Galaxy | 67 |
| M70 NGC 6681 | Sgr | 18 | 42.2 | –32 | 18 | 8.0 | Globular Cluster | 89 |
| M110 NGC 205 | And | 00 | 40.4 | +41 | 41 | 8.1 | Galaxy | 53 |
| M32 NGC 221 | And | 00 | 42.7 | +40 | 52 | 8.1 | Galaxy | 53 |
| M107 NGC 6171 | Oph | 16 | 32.5 | –13 | 03 | 8.1 | Globular Cluster | 87 |
| NGC 6543 Cat's Eye Nebula | Dra | 17 | 58.6 | +66 | 38 | 8.1 | Planetary Nebula | 153 |
| NGC 6572 | Oph | 18 | 12.1 | +06 | 51 | 8.1 | Planetary Nebula | 155 |
| M94 NGC 4736 | CVn | 12 | 50.9 | +41 | 07 | 8.2 | Galaxy | 62 |
| NGC 6712 | Sct | 18 | 53.1 | –08 | 42 | 8.2 | Globular Cluster | 91 |
| M56 NGC 6779 | Lyr | 19 | 16.6 | +30 | 11 | 8.3 | Globular Cluster | 89 |
| M71 NGC 6838 | Sge | 19 | 53.8 | +18 | 47 | 8.3 | Globular Cluster | 90 |
| NGC 7662 Blue Snowball | And | 23 | 25.9 | +42 | 33 | 8.3 | Planetary Nebula | 156 |
| NGC 7009 Saturn Nebula | Aqr | 21 | 04.2 | –11 | 22 | 8.3 | Planetary Nebula | 157 |
| IC 342 | Cam | 3 | 46.8 | +68 | 06 | 8.4 | Galaxy | 57 |
| M1 NGC 1952 Crab Nebula | Tau | 05 | 34.5 | +22 | 01 | 8.4 | Supernova Remnant | 163 |
| M82 NGC 3034 | UMa | 09 | 55.8 | +69 | 41 | 8.4 | Galaxy | 56 |
| M49 NGC 4472 | Vir | 12 | 29.8 | +08 | 00 | 8.4 | Galaxy | 68 |
| M51 NGC 5194 Whirlpool Galaxy | CVn | 13 | 29.9 | +47 | 12 | 8.4 | Galaxy | 60 |
| NGC 2403 | Cam | 07 | 36.9 | +65 | 36 | 8.5 | Galaxy | 57 |
| M64 NGC 4826 Black Eye Galaxy | Com | 12 | 56.7 | +21 | 41 | 8.5 | Galaxy | 62 |
| M75 NGC 6864 | Sgr | 20 | 06.1 | –21 | 55 | 8.5 | Globular Cluster | 84 |
| NGC 2158 | Gem | 06 | 07.5 | +23 | 18 | 8.6 | Open Cluster | 122 |
| M87 NGC 4486 | Vir | 12 | 30.8 | +12 | 24 | 8.6 | Galaxy | 66 |

| Designation/Name | Constellation | RA h | m | Dec ° | ' | Mag. | Type | Page |
|---|---|---|---|---|---|---|---|---|
| M63 NGC 5055 Sunflower Galaxy | CVn | 13 | 15.8 | +42 | 02 | 8.6 | Galaxy | 61 |
| NGC 6934 | Del | 20 | 34.2 | +07 | 24 | 8.7 | Globular Cluster | 91 |
| M57 NGC 6720 Ring Nebula | Lyr | 18 | 53.6 | +33 | 02 | 8.8 | Planetary Nebula | 150 |
| M60 NGC 4649 | Vir | 12 | 43.7 | +11 | 33 | 8.8 | Galaxy | 63 |
| NGC 6826 Blinking Planetary | Cyg | 19 | 44.8 | +50 | 31 | 8.8 | Planetary Nebula | 155 |
| M77 NGC 1068 | Cet | 02 | 42.7 | −00 | 01 | 8.9 | Galaxy | 69 |
| NGC 3115 Spindle Galaxy | Sex | 10 | 05.2 | −07 | 43 | 8.9 | Galaxy | 70 |
| M66 NGC 3627 | Leo | 11 | 20.2 | +12 | 59 | 8.9 | Galaxy | 58 |
| M86 NGC 4406 | Vir | 12 | 26.2 | +12 | 57 | 8.9 | Galaxy | 66 |
| NGC 2903 | Leo | 09 | 32.2 | +21 | 30 | 9.0 | Galaxy | 60 |
| NGC 1499 California Nebula | Per | 04 | 00.7 | +36 | 37 | 9 | Emission Nebula | 103 |
| M85 NGC 4382 | Com | 12 | 25.4 | +18 | 11 | 9.1 | Galaxy | 64 |
| M84 NGC 4374 | Vir | 12 | 25.1 | +12 | 53 | 9.1 | Galaxy | 66 |
| NGC 2392 Eskimo Nebula | Gem | 07 | 29.2 | +20 | 55 | 9.2 | Planetary Nebula | 151 |
| M96 NGC 3368 | Leo | 10 | 46.8 | +10 | 49 | 9.2 | Galaxy | 59 |
| M105 NGC 3379 | Leo | 10 | 47.8 | +12 | 35 | 9.3 | Galaxy | 59 |
| M65 NGC 3623 | Leo | 11 | 18.9 | +13 | 05 | 9.3 | Galaxy | 58 |
| M100 NGC 4321 | Com | 12 | 22.6 | +15 | 47 | 9.3 | Galaxy | 64 |
| M72 NGC 6981 | Aqr | 20 | 53.5 | −12 | 32 | 9.3 | Globular Cluster | 83 |
| M74 NGC 628 | Psc | 01 | 36.7 | +15 | 47 | 9.4 | Galaxy | 69 |
| NGC 2440 | Pup | 07 | 41.9 | −18 | 13 | 9.4 | Planetary Nebula | 158 |
| NGC 3628 | Leo | 11 | 20.3 | +13 | 36 | 9.5 | Galaxy | 58 |
| M90 NGC 4569 | Vir | 12 | 36.8 | +13 | 10 | 9.5 | Galaxy | 63 |
| NGC 1535 | Eri | 04 | 14.2 | −12 | 44 | 9.6 | Planetary Nebula | 158 |
| M88 NGC 4501 | Com | 12 | 32.0 | +14 | 25 | 9.6 | Galaxy | 64 |
| NGC 4565 | Com | 12 | 36.3 | +25 | 59 | 9.6 | Galaxy | 71 |
| M59 NGC 4621 | Vir | 12 | 42.0 | +11 | 39 | 9.6 | Galaxy | 63 |
| M95 NGC 3351 | Leo | 10 | 44.0 | +11 | 42 | 9.7 | Galaxy | 59 |
| M61 NGC 4303 | Vir | 12 | 21.9 | +04 | 28 | 9.7 | Galaxy | 68 |
| M58 NGC 4579 | Vir | 12 | 37.7 | +11 | 49 | 9.7 | Galaxy | 63 |
| M89 NGC 4552 | Vir | 12 | 35.7 | +12 | 33 | 9.8 | Galaxy | 63 |
| NGC 891 | And | 02 | 22.6 | +42 | 21 | 9.9 | Galaxy | 71 |
| M97 NGC 3587 Owl Nebula | UMa | 11 | 14.8 | +55 | 01 | 9.9 | Planetary Nebula | 152 |
| M99 NGC 4254 | Com | 12 | 18.8 | +14 | 25 | 9.9 | Galaxy | 64 |
| NGC 2237–2239, NGC 2246 Rosette Nebula | Mon | 06 | 32.3 | +05 | 03 | 10 | Emission Nebula | 104 |
| NGC 2261 Hubble's Variable Nebula | Mon | 06 | 39.2 | +08 | 44 | c.10 | Reflection Nebula | 107 |
| M76 NGC 650/651 Little Dumbbell | Per | 01 | 42.4 | +51 | 34 | 10.1 | Planetary Nebula | 153 |

| Designation/Name | Constellation | RA h | m | Dec ° | ' | Mag. | Type | Page |
|---|---|---|---|---|---|---|---|---|
| M98 NGC 4192 | Com | 12 | 13.8 | +14 | 54 | 10.1 | Galaxy | 64 |
| M91 NGC 4548 | Com | 12 | 35.4 | +14 | 30 | 10.2 | Galaxy | 64 |
| NGC 2419 Intergalactic Tramp | Lyn | 07 | 38.1 | +38 | 53 | 10.3 | Globular Cluster | 93 |
| NGC 6891 | Del | 20 | 15.2 | +12 | 42 | 10.5 | Planetary Nebula | 159 |
| NGC 7006 | Del | 21 | 01.4 | +16 | 12 | 10.5 | Globular Cluster | 91 |
| IC 3568 | Cam | 12 | 32.9 | +82 | 33 | 10.6 | Planetary Nebula | 159 |
| NGC 7026 | Cyg | 21 | 06.3 | +47 | 51 | 10.9 | Planetary Nebula | 156 |
| NGC 2438 | Pup | 07 | 41.8 | −14 | 44 | 11.0 | Planetary Nebula | 158 |

## Objects by Season

The Earth's ever-changing orbital position places different objects on view at different times of year. This table gives a guide to those best-presented by season, and will help the reader to plan observing sessions.

| Objects by Season | | | | | |
|---|---|---|---|---|---|
| Designation/Name | Constellation | RA h | m | Dec ° | ' |
| 47 Tucanae NGC 104 | Tuc | 00 | 24.1 | −72 | 05 |
| M110 NGC 205 | And | 00 | 40.4 | +41 | 41 |
| M31 NGC 224 Andromeda Galaxy | And | 00 | 42.7 | +41 | 16 |
| M32 NGC 221 | And | 00 | 42.7 | +40 | 52 |
| NGC 253 Silver Coin | Scl | 00 | 47.6 | −25 | 17 |
| Small Magellanic Cloud | Tuc | 00 | 52.7 | −72 | 50 |
| NGC 457 | Cas | 01 | 19.1 | +58 | 20 |
| M103 NGC 581 | Cas | 01 | 33.2 | +60 | 52 |
| M33 NGC 598 | Tri | 01 | 33.9 | +30 | 39 |
| M74 NGC 628 | Psc | 01 | 36.7 | +15 | 47 |
| M76 NGC 650/651 Little Dumbbell | Per | 01 | 42.4 | +51 | 34 |
| γ Ari | Ari | 01 | 53.5 | +19 | 18 |
| NGC 752 | And | 01 | 57.8 | +37 | 41 |
| α Psc | Psc | 02 | 02.0 | +02 | 46 |
| NGC 869 | Per | 02 | 19.0 | +57 | 09 |
| NGC 884 | Per | 02 | 22.4 | +57 | 07 |
| M34 NGC 1039 | Per | 02 | 42.0 | +42 | 47 |
| ε Ari | Ari | 02 | 59.2 | +21 | 20 |
| γ And Almach | And | 02 | 03.9 | +42 | 20 |
| NGC 891 | And | 02 | 22.6 | +42 | 21 |
| M77 NGC 1068 | Cet | 02 | 42.7 | −00 | 01 |
| NGC 1435 Merope Nebula | Tau | 03 | 46.1 | +25 | 47 |
| IC 342 | Cam | 03 | 46.8 | +68 | 06 |
| M45 Pleiades | Tau | 03 | 47.0 | +24 | 07 |
| NGC 1499 California Nebula | Per | 04 | 00.7 | +36 | 37 |
| Kemble's Cascade | Cam | 04 | 10 | +63 | |
| NGC 1528 | Per | 04 | 15.4 | +51 | 14 |
| NGC 1535 | Eri | 04 | 14.2 | −12 | 44 |

## Objects by Season

| Mag. | Type | Page | Viewing period |
|---|---|---|---|
| 4.0 | Globular Cluster | 77 | All year (Southern Hemisphere) South Circumpolar |
| 8.1 | Galaxy | 53 | September to February |
| 3.4 | Galaxy | 53 | September to February |
| 8.1 | Galaxy | 53 | September to February |
| 7.8 | Galaxy | 70 | September to February |
| 2.3 | Galaxy | 51 | All year (Southern Hemisphere) South Circumpolar |
| 6.4 | Open Cluster | 129 | July to March |
| 7.4 | Open Cluster | 129 | July to March |
| 5.7 | Galaxy | 55 | September to February |
| 9.4 | Galaxy | 69 | September to February |
| 10.1 | Planetary Nebula | 153 | July to March |
| | Double Star | 138 | July to March |
| 5.7 | Open Cluster | 130 | July to March |
| | Double Star | 145 | August to February |
| 5.3 | Open Cluster | 118 | July to April |
| 6.1 | Open Cluster | 118 | July to April |
| 5.2 | Open Cluster | 120 | July to March |
| | Double Star | 145 | August to February |
| | Double Star | 140 | July to February |
| 9.9 | Galaxy | 71 | September to February |
| 8.9 | Galaxy | 69 | September to February |
| | Reflection Nebula | 107 | August to February |
| 8.4 | Galaxy | 57 | August to February |
| 1.2 | Open Cluster | 114 | August to February |
| 9 | Emission Nebula | 103 | August to February |
| | Asterism | 132 | August to February |
| 6.4 | Open Cluster | 120 | August to February |
| 9.6 | Planetary Nebula | 158 | September to February |

| Designation/Name | Constellation | RA h | m | Dec ° | ' |
|---|---|---|---|---|---|
| o² Eri | Eri | 04 | 15.3 | −07 | 39 |
| Hyades | Tau | 04 | 27 | +16 | |
| Large Magellanic Cloud | Dor | 05 | 23.6 | −69 | 45 |
| M79 NGC 1904 | Lep | 05 | 24.5 | −24 | 33 |
| M38 NGC 1912 | Aur | 05 | 28.7 | +39 | 50 |
| M1 NGC 1952 Crab Nebula | Tau | 05 | 34.5 | +22 | 01 |
| θ¹ Ori Trapezium | Ori | 05 | 35.3 | −05 | 23 |
| M42 NGC 1976 Orion Nebula | Ori | 05 | 35.4 | −05 | 27 |
| M43 NGC 1982 Ori | Ori | 05 | 35.6 | −05 | 16 |
| M36 NGC 1960 | Aur | 05 | 36.1 | +34 | 08 |
| NGC 2070 Tarantula Nebula | Dor | 05 | 38.7 | −69 | 06 |
| ζ Ori (Alnitak) | Ori | 05 | 40.8 | −01 | 57 |
| M78 NGC 2067/206 | Ori | 05 | 46.7 | +00 | 03 |
| M37 NGC 2099 | Aur | 05 | 52.4 | +32 | 33 |
| NGC 2158 | Gem | 06 | 07.5 | +23 | 18 |
| M35 NGC 2168 | Gem | 06 | 08.9 | +24 | 20 |
| β Mon | Mon | 06 | 28.8 | −07 | 02 |
| NGC 2237–2239, NGC 2246 Rosette Nebula | Mon | 06 | 32.3 | +05 | 03 |
| NGC 2244 | Mon | 06 | 32.4 | +04 | 52 |
| NGC 2261 Hubble's Variable Nebula | Mon | 06 | 39.2 | +08 | 44 |
| M41 NGC 2287 | CMa | 06 | 47.0 | −20 | 44 |
| M50 NGC 2323 | Mon | 07 | 03.2 | −08 | 20 |
| NGC 2362 Tau CMa cluster | CMa | 07 | 18.8 | −26 | 47 |
| NGC 2392 Eskimo Nebula | Gem | 07 | 29.2 | +20 | 55 |
| α Gem (Castor) | Gem | 07 | 34.6 | +31 | 53 |
| M47 NGC 2422 | Pup | 07 | 36.6 | −14 | 30 |
| NGC 2403 | Cam | 07 | 36.9 | +65 | 36 |
| NGC 2419 Intergalactic Tramp | Lyn | 07 | 38.1 | +38 | 53 |
| M46 NGC 2437 | Pup | 07 | 41.8 | −14 | 49 |
| NGC 2438 | Pup | 07 | 41.8 | − 14 | 44 |
| NGC 2440 | Pup | 07 | 41.9 | −18 | 13 |
| M93 NGC 2447 | Pup | 07 | 44.6 | −23 | 52 |
| M48 NGC 2458 | Hya | 08 | 13.8 | −05 | 48 |
| M44 NGC 2632 Praesepe | Cnc | 08 | 40.1 | +19 | 59 |
| IC 2391 | Vel | 08 | 40.2 | −53 | 04 |
| ι Cnc | Cnc | 08 | 46.7 | +28 | 46 |
| M67 NGC 2682 | Cnc | 08 | 50.4 | +11 | 49 |
| NGC 2736 | Vel | 09 | 00.4 | −45 | 54 |
| NGC 2903 | Leo | 09 | 32.2 | +21 | 30 |
| M81 NGC 3031 | UMa | 09 | 55.6 | +69 | 04 |
| M82 NGC 3034 | UMa | 09 | 55.8 | +69 | 41 |
| NGC 3115 Spindle Galaxy | Sex | 10 | 05.2 | −07 | 43 |
| γ Leo (Algieba) | Leo | 10 | 20.0 | +19 | 51 |
| NGC 3242 Ghost of Jupiter | Hya | 10 | 24.8 | −18 | 38 |
| IC 2602 Southern Pleiades | Car | 10 | 43.2 | −64 | 24 |
| NGC 3372 Eta Carinae Nebula | Car | 10 | 43.8 | −59 | 52 |

| Mag. | Type | Page | Viewing period |
|------|------|------|----------------|
|  | Double Star | 137 | September to February |
| 0.5 | Open Cluster | 116 | August to April |
| 0.1 | Galaxy | 51 | All year (Southern Hemisphere) South Circumpolar |
| 7.8 | Globular Cluster | 87 | October to February |
| 6.4 | Open Cluster | 120 | September to April |
| 8.4 | Supernova Remnant | 163 | September to March |
|  | Multiple Star | 141 | September to February |
| 4.0 | Emission Nebula | 99 | September to February |
| 5.0 | Emission Nebula | 99 | September to February |
| 6.0 | Open Cluster | 120 | September to April |
| 8.0(?) | Emission Nebula | 103 | All year (Southern Hemisphere) South Circumpolar |
|  | Double Star | 143 | September to February |
| 8 | Reflection Nebula | 105 | September to February |
| 5.6 | Open Cluster | 121 | September to April |
| 8.6 | Open Cluster | 122 | September to April |
| 5.1 | Open Cluster | 122 | September to April |
|  | Triple Star | 142 | September to March |
| 10 | Emission Nebula | 104 | September to March |
| 4.8 | Open Cluster | 123 | September to March |
| c.10 | Reflection Nebula | 107 | September to March |
| 4.5 | Open Cluster | 123 | October to February |
| 5.9 | Open Cluster | 122 | October to February |
| 4.1 | Open Cluster | 124 | October to February |
| 9.2 | Planetary Nebula | 151 | September to April |
|  | Double Star | 143 | September to April |
| 4.4 | Open Cluster | 124 | October to April |
| 8.5 | Galaxy | 57 | August to May |
| 10.3 | Globular Cluster | 93 | September to April |
| 6.1 | Open Cluster | 124 | October to April |
| 11. | Planetary Nebula | 158 | October to April |
| 9.4 | Planetary Nebula | 158 | October to April |
| 6.2 | Open Cluster | 125 | October to April |
| 5.8 | Open Cluster | 125 | October to April |
| 3.1 | Open Cluster | 117 | October to April |
| 2.5 | Open Cluster | 126 | October to May (Southern Hemisphere) |
|  | Double Star | 140 | October to April |
| 6.9 | Open Cluster | 118 | October to April |
|  | Supernova Remnant | 167 | October to May (Southern Hemisphere) |
| 9.0 | Galaxy | 60 | October to May |
| 6.9 | Galaxy | 56 | October to May |
| 8.4 | Galaxy | 56 | October to May |
| 8.9 | Galaxy | 70 | November to May |
|  | Double Star | 144 | October to May |
| 7.8 | Planetary Nebula | 154 | November to April |
| 1.9 | Open Cluster | 117 | October to July (Southern Hemisphere) |
| 1.0 | Emission Nebula | 98 | October to July (Southern Hemisphere) |

| Designation/Name | Constellation | RA h | m | Dec ° | ′ |
|---|---|---|---|---|---|
| M95 NGC 3351 | Leo | 10 | 44.0 | +11 | 42 |
| M96 NGC 3368 | Leo | 10 | 46.8 | +10 | 49 |
| M105 NGC 3379 | Leo | 10 | 47.8 | +12 | 35 |
| M97 NGC 3587 The Owl Nebula | UMa | 11 | 14.8 | +55 | 01 |
| ξ Ursae Majoris | UMa | 11 | 18.2 | +31 | 32 |
| M65 NGC 3623 | Leo | 11 | 18.9 | +13 | 05 |
| M66 NGC 3627 | Leo | 11 | 20.2 | +12 | 59 |
| NGC 3628 | Leo | 11 | 20.3 | +13 | 36 |
| M98 NGC 4192 | Com | 12 | 13.8 | +14 | 54 |
| M99 NGC 4254 | Com | 12 | 18.8 | +14 | 25 |
| M61 NGC 4303 | Vir | 12 | 21.9 | +04 | 28 |
| M40 | UMa | 12 | 22.4 | +58 | 05 |
| M100 NGC 4321 | Com | 12 | 22.6 | +15 | 47 |
| Mel 111 | Com | 12 | 25 | +26 | |
| M85 NGC 4382 | Com | 12 | 25.4 | +18 | 11 |
| α Cru (Acrux) | Cru | 12 | 26.6 | −63 | 06 |
| M84 NGC 4374 | Vir | 12 | 25.1 | +12 | 53 |
| M86 NGC 4406 | Vir | 12 | 26.2 | +12 | 57 |
| M49 NGC 4472 | Vir | 12 | 29.8 | +08 | 00 |
| M87 NGC 4486 | Vir | 12 | 30.8 | +12 | 24 |
| M88 NGC 4501 | Com | 12 | 32.0 | +14 | 25 |
| IC 3568 | Cam | 12 | 32.9 | +82 | 33 |
| M91 NGC 4548 | Com | 12 | 35.4 | +14 | 30 |
| M89 NGC 4552 | Vir | 12 | 35.7 | +12 | 33 |
| NGC 4565 | Com | 12 | 36.3 | +25 | 59 |
| M90 NGC 4569 | Vir | 12 | 36.8 | +13 | 10 |
| M58 NGC 4579 | Vir | 12 | 37.7 | +11 | 49 |
| M68 NGC 4590 | Hya | 12 | 39.5 | −26 | 45 |
| M104 NGC 4594 | Vir | 12 | 40.0 | −11 | 37 |
| Sombrero Galaxy | | | | | |
| M59 NGC 4621 | Vir | 12 | 42.0 | +11 | 39 |
| M60 NGC 4649 | Vir | 12 | 43.7 | +11 | 33 |
| M64 NGC 4826 Black Eye Galaxy | Com | 12 | 56.7 | +21 | 41 |
| NGC 4833 | Mus | 12 | 59.6 | −70 | 53 |
| M53 NGC 5024 | Com | 13 | 12.9m | +18 | 10 |
| NGC 5139 Omega Centauri | Cen | 13 | 26.8 | −47 | 29 |
| M94 NGC 4736 | CVn | 12 | 50.9 | +41 | 07 |
| α CVn (Cor Caroli) | CVn | 12 | 56.0 | +38 | 19 |
| M63 NGC 5055 Sunflower Galaxy | CVn | 13 | 15.8 | +42 | 02 |
| ζ UMa (Mizar) | UMa | 13 | 23.9 | +54 | 56 |
| NGC 5128 | Cen | 13 | 25.5 | −43 | 01 |
| M51 NGC 5194 Whirlpool Galaxy | CVn | 13 | 29.9 | +47 | 12 |
| M83 NGC 5236 | Hya | 13 | 37.0 | −29 | 52 |
| M3 NGC 5272 | CVn | 13 | 42.2 | +28 | 23 |
| M101 NGC 5457 | UMa | 14 | 03.2 | +54 | 21 |
| α Cen (Rigil Kentaurus) | Cen | 14 | 39.6 | −60 | 50 |

| Mag. | Type | Page | Viewing period |
|------|------|------|----------------|
| 9.7 | Galaxy | 59 | November to May |
| 9.2 | Galaxy | 59 | November to May |
| 9.3 | Galaxy | 59 | November to May |
| 9.9 | Planetary Nebula | 152 | October to June |
| | Double Star | 143 | October to June |
| 9.3 | Galaxy | 58 | November to May |
| 8.9 | Galaxy | 58 | November to May |
| 9.5 | Galaxy | 58 | November to May |
| 10.1 | Galaxy | 64 | December to June |
| 9.9 | Galaxy | 64 | December to June |
| 9.7 | Galaxy | 68 | December to June |
| | Asterism | 131 | November to May |
| 9.3 | Galaxy | 64 | December to June |
| 1.8 | Open Cluster | 116 | November to July |
| 9.1 | Galaxy | 64 | December to June |
| | Double Star | 145 | November to August (Southern Hemisphere) |
| 9.1 | Galaxy | 66 | December to June |
| 8.9 | Galaxy | 66 | December to June |
| 8.4 | Galaxy | 68 | December to June |
| 8.6 | Galaxy | 66 | December to June |
| 9.6 | Galaxy | 64 | December to June |
| 10.6 | Planetary Nebula | 159 | November to July |
| 10.2 | Galaxy | 64 | December to June |
| 9.8 | Galaxy | 63 | December to June |
| 9.6 | Galaxy | 71 | December to June |
| 9.5 | Galaxy | 63 | December to June |
| 9.7 | Galaxy | 63 | December to June |
| 7.7 | Globular Cluster | 89 | December to June |
| 8.0 | Galaxy | 67 | December to June |
| | | | |
| 9.6 | Galaxy | 63 | December to June |
| 8.8 | Galaxy | 63 | December to June |
| 8.5 | Galaxy | 62 | December to June |
| 6.9 | Globular Cluster | 87 | December to September (Southern Hemisphere) |
| 7.5 | Globular Cluster | 82 | December to June |
| 3.5 | Globular Cluster | 76 | December to July (Southern Hemisphere) |
| 8.2 | Galaxy | 62 | December to July |
| | Double Star | 137 | December to July |
| 8.6 | Galaxy | 61 | December to July |
| | Double Star | 136 | All year (North circumpolar) |
| 6.7 | Galaxy | 58 | December to August (Southern Hemisphere) |
| 8.4 | Galaxy | 60 | November to July |
| 7.6 | Galaxy | 70 | January to May |
| 5.9 | Globular Cluster | 81 | December to June |
| 7.9 | Galaxy | 61 | December to June |
| | Double Star | 137 | December to September (Southern Hemisphere) |

| Designation/Name | Constellation | RA h | m | Dec ° | ' |
|---|---|---|---|---|---|
| ε Boo | Boo | 14 | 45.0 | +27 | 04 |
| M5 NGC 5904 | Ser | 15 | 18.6 | +02 | 05 |
| NGC 4755 Jewel Box | Cru | 15 | 53.6 | −60 | 20 |
| β Sco | Sco | 16 | 05.4 | −19 | 48 |
| ν Sco | Sco | 16 | 12.0 | −19 | 28 |
| M80 NGC 6093 | Sco | 16 | 17.0 | −22 | 59 |
| M4 NGC 6121 | Sco | 16 | 23.6 | −26 | 32 |
| α Sco (Antares) | Sco | 16 | 29.4 | −26 | 26 |
| M107 NGC 6171 | Oph | 16 | 32.5 | −13 | 03 |
| M13 NGC 6205 | Her | 16 | 41.7 | +36 | 28 |
| M12 NGC 6218 | Oph | 16 | 47.2 | −01 | 57 |
| M10 NGC 6254 | Oph | 16 | 57.1 | −04 | 06 |
| M62 NGC 6266 | Oph | 17 | 01.2 | −30 | 07 |
| M19 NGC 6273 | Oph | 17 | 02.6 | −26 | 16 |
| α Her (Rasalgethi) | Her | 17 | 14.6 | +14 | 23 |
| M92 NGC 6341 | Her | 17 | 17.1 | +43 | 08 |
| B64 | Oph | 17 | 17.2 | −18 | 33 |
| M9 NGC 6333 | Oph | 17 | 19.2 | −18 | 31 |
| B59 | Oph | 17 | 21.0 | −27 | 00 |
| B72 | Oph | 17 | 23.5 | −23 | 28 |
| ν Dra | Dra | 17 | 32.2 | +55 | 10 |
| B78 | Oph | 17 | 33.0 | −26 | 00 |
| M14 NGC 6402 | Oph | 17 | 37.6 | −03 | 15 |
| M6 NGC 6405 Butterfly Cluster | Sco | 17 | 40.1 | −32 | 13 |
| NGC 6397 | Ara | 17 | 40.7 | −53 | 40 |
| M7 NGC 6475 | Sco | 17 | 53.9 | −34 | 49 |
| IC 4665 | Oph | 17 | 46 | +05 | 43 |
| NGC 6543 Cat's Eye Nebula | Dra | 17 | 58.6 | +66 | 38 |
| M20 NGC 6514 Trifid Nebula | Sgr | 18 | 02.3 | −23 | 02 |
| M8 NGC 6523 Lagoon Nebula | Sgr | 18 | 03.8 | −24 | 21 |
| NGC 6572 | Oph | 18 | 12.1 | +06 | 51 |
| M24 Small Sagittarius Star Cloud | Sgr | 18 | 16.5 | −18 | 50 |
| B92 | Sgr | 18 | 16.9 | −18 | 02 |
| M16 IC 4703/NGC 6611 Eagle Nebula | Ser | 18 | 18.8 | −13 | 47 |
| M17 NGC 6618 Swan Nebula | Sgr | 18 | 20.8 | −16 | 11 |
| M28 NGC 6626 | Sgr | 18 | 24.5 | −24 | 52 |
| M69 NGC 6637 | Sgr | 18 | 31.4 | −32 | 21 |
| M25 IC 4725 | Sgr | 18 | 31.6 | −19 | 15 |
| M22 NGC 6656 | Sgr | 18 | 36.4 | −23 | 54 |
| M70 NGC 6681 | Sgr | 18 | 42.2 | −32 | 18 |
| ε Lyr Double Double | Lyr | 18 | 44.3 | +39 | 40 |
| ζ Lyr | Lyr | 18 | 44.8 | +37 | 36 |
| M11 NGC 6705 Wild Duck | Sct | 18 | 51.1 | −06 | 16 |
| M57 NGC 6720 Ring Nebula | Lyr | 18 | 53.6 | +33 | 02 |
| NGC 6712 | Sct | 18 | 53.1 | −08 | 42 |
| M54 NGC 6715 | Sgr | 18 | 55.1 | −30 | 29 |

| Mag. | Type | Page | Viewing period |
|------|------|------|----------------|
| | Double Star | 144 | January to August |
| 5.7 | Globular Cluster | 80 | February to August |
| 4.2 | Open Cluster | 126 | December to August (Southern Hemisphere) |
| | Double Star | 139 | March to August |
| | Multiple Star | 143 | March to August |
| 7.3 | Globular Cluster | 81 | March to August |
| 5.8 | Globular Cluster | 81 | March to August |
| | Double Star | 145 | March to August |
| 8.1 | Globular Cluster | 87 | March to August |
| 5.7 | Globular Cluster | 79 | February to September |
| 6.8 | Globular Cluster | 85 | March to August |
| 6.6 | Globular Cluster | 85 | March to August |
| 6.7 | lobular Cluster | 86 | March to August |
| 6.7 | Globular Cluster | 85 | March to August |
| | Double Star | 141 | March to September |
| 6.5 | Globular Cluster | 80 | February to October |
| | Dark Nebula | 110 | March to August |
| 7.6 | Globular Cluster | 86 | March to August |
| | Dark Nebula | 108 | March to August |
| | Dark Nebula | 110 | March to August |
| | Double Star | 139 | February to October |
| | Dark Nebula | 108 | March to August |
| 7.6 | Globular Cluster | 87 | March to August |
| 4.2 | Open Cluster | 126 | April to August |
| 5.9 | Globular Cluster | 81 | April to August (Southern Hemisphere) |
| 3.3 | Open Cluster | 127 | April to August |
| 4.2 | Open Cluster | 126 | March to September |
| 8.1 | Planetary Nebula | 153 | February to November |
| 6.3 | Emission Nebula | 101 | April to September |
| 6.0 | Emission Nebula | 100 | April to September |
| 8.1 | Planetary Nebula | 155 | March to October |
| 4.6 | Open Cluster | 127 | April to September |
| | Dark Nebula | 110 | April to September |
| 6.0 | Emission Nebula | 102 | April to September |
| 6.0 | Emission Nebula | 102 | April to September |
| 6.8 | Globular Cluster | 78 | April to September |
| 7.6 | Globular Cluster | 89 | April to September |
| 4.6 | Open Cluster | 127 | April to September |
| 5.1 | Globular Cluster | 127 | April to September |
| 8.0 | Globular Cluster | 89 | April to September |
| | Multiple Star | 142 | March to November |
| | Double Star | 138 | March to November |
| 5.8 | Open Cluster | 128 | April to October |
| 8.8 | Planetary Nebula | 150 | March to November |
| 8.2 | Globular Cluster | 91 | April to October |
| 7.6 | Globular Cluster | 90 | May to September |

| Designation/Name | Constellation | RA h | m | Dec ° | ′ |
|---|---|---|---|---|---|
| θ Ser | Ser | 18 | 56.2 | +04 | 12 |
| B133 | Aql | 19 | 06.1 | −06 | 50 |
| NGC 6752 | Pav | 19 | 10.9 | −59 | 59 |
| M56 NGC 6779 | Lyr | 19 | 16.6 | +30 | 11 |
| Cr 399 Coathanger | Vul | 19 | 25.4 | +20 | 11 |
| β Cyg (Albireo) | Cyg | 19 | 30.7 | +27 | 58 |
| M55 NGC 6809 | Sgr | 19 | 40.0 | −30 | 58 |
| B142 | Aql | 19 | 40.7 | +10 | 57 |
| NGC 6826 Blinking Planetary | Cyg | 19 | 44.8 | +50 | 31 |
| M71 NGC 6838 | Sge | 19 | 53.8 | +18 | 47 |
| M27 NGC 6583 Dumbbell Nebula | Vul | 19 | 59.6 | +22 | 43 |
| M75 NGC 6864 | Sgr | 20 | 06.1 | −21 | 55 |
| o¹ Cyg | Cyg | 20 | 13.6 | +46 | 44 |
| 61 Cyg | Cyg | 20 | 13.6 | +38 | 45 |
| NGC 6891 | Del | 20 | 15.2 | +12 | 42 |
| β Cap | Cap | 20 | 21.0 | −14 | 47 |
| M29 NGC 6913 | Cyg | 20 | 23.9 | +38 | 32 |
| NGC 6934 | Del | 20 | 34.2 | +07 | 24 |
| γ Del | Del | 20 | 46.7 | +16 | 07 |
| Veil Nebula | Cyg | 20 | 50 | +32 | |
| M72 NGC 6981 | Aqr | 20 | 53.5 | −12 | 32 |
| NGC 7000 North America Nebula | Cyg | 20 | 58.8 | +44 | 20 |
| NGC 7026 | Cyg | 21 | 06.3 | +47 | 51 |
| M39 NGC 7092 | Cyg | 21 | 32.2 | +48 | 26 |
| M73 NGC 6994 | Aqr | 20 | 59.0 | −12 | 58 |
| NGC 7006 | Del | 21 | 01.4 | +16 | 12 |
| NGC 7009 Saturn Nebula | Aqr | 21 | 04.2 | −11 | 22 |
| M15 NGC 7078 Peg | Peg | 21 | 30.0 | +12 | 10 |
| M2 NGC 7089 | Aqr | 21 | 33.5 | −00 | 49 |
| M30 NGC 7099 | Cap | 21 | 40.4 | −23 | 11 |
| ζ Aqr | Aqr | 22 | 28.8 | −00 | 01 |
| NGC 7293 Helix Nebula | Aqr | 22 | 29.6 | −20 | 48 |
| M52 NGC 7654 | Cas | 23 | 24.2 | +61 | 35 |
| NGC 7662 Blue Snowball | And | 23 | 25.9 | +42 | 33 |

| Mag. | Type | Page | Viewing period |
|------|------|------|----------------|
| | Double Star | 138 | April to November |
| | Dark Nebula | 110 | May to October |
| 5.4 | Globular Cluster | 78 | March to November (Southern Hemisphere) |
| 8.3 | Globular Cluster | 89 | March to November |
| | Asterism | 131 | March to November |
| | Double Star | 139 | March to November |
| 6.4 | Globular Cluster | 84 | May to September |
| | Dark Nebula | 110 | May to October |
| 8.8 | Planetary Nebula | 155 | March to November |
| 8.3 | Globular Cluster | 90 | March to November |
| 7.3 | Planetary Nebula | 149 | March to November |
| 8.5 | Globular Cluster | 84 | May to October |
| | Multiple Star | 140 | March to November |
| | Double Star | 138 | March to November |
| 10.5 | Planetary Nebula | 159 | April to November |
| | Double Star | 139 | May to November |
| 6.6 | Open Cluster | 130 | March to December |
| 8.7 | Globular Cluster | 91 | April to November |
| | Double Star | 141 | April to November |
| | Supernova Remnant | 164 | April to November |
| 9.3 | Globular Cluster | 83 | May to November |
| 5? | Emission Nebula | 104 | March to November |
| 10.9 | Planetary Nebula | 156 | March to November |
| 5.0 | Open Cluster | 130 | March to November |
| | Asterism | 131 | May to November |
| 10.5 | Globular Cluster | 91 | April to November |
| 8.3 | Planetary Nebula | 157 | May to November |
| 6.0 | Globular Cluster | 83 | May to December |
| 6.4 | Globular Cluster | 83 | May to December |
| 7.3 | Globular Cluster | 84 | May to October |
| | Double Star | 145 | May to December |
| 7.3 | Planetary Nebula | 157 | May to December |
| 6.3 | Open Cluster | 128 | July to March |
| 8.3 | Planetary Nebula | 156 | June to February |

## Books

Burnham, R., *Burnham's Celestial Handbook* (3 volumes). Dover (1978)

Consolmagno, G., and Davis, D.M., *Turn Left at Orion*, 3rd edition. Cambridge University Press (2000)

Houston, W.S., *Deep Sky Wonders* (edited by S.J. O'Meara). Sky Publishing/Cambridge University Press (1999)

Kepple, G.R., and Sanner, G.W., *The Night Sky Observer's Guide* (2 volumes). Willmann-Bell (1998)

McRobert, A.M., *Star-Hopping for Backyard Astronomers*. Sky Publishing (1993)

Mallas, J.H., and Kreimer, E., *The Messier Album*. Sky Publishing/Cambridge University Press (1979)

Moore, P. (general editor), *Astronomy Encyclopedia*. Philip's (2002)

O'Meara, S.J., *Deep Sky Companions: The Messier Objects*. Sky Publishing/Cambridge University Press (1999)

Pennington, H., *The Year-Round Messier Marathon*. Willmann-Bell (1997)

Ridpath, I. (editor), *Norton's Star Atlas*, 20th edition. Pi Press (2003)

Scagell, R., *Stargazing with a Telescope*. Philip's (2000)

Sinnott, R.W., *NGC 2000.0 (The Complete New General Catalogue and Index Catalogues of Nebulae and Star Clusters by J.L.E. Dreyer)*. Sky Publishing/Cambridge University Press (1988)

Smyth, W.H., *The Bedford Catalogue* (from *A Cycle of Celestial Objects*). Willmann-Bell (1986)

Tirion, W., Rappaport, B., and Remalkus, P., *Uranometria 2000.0*, 2nd edition (2 volumes). Willmann-Bell (2001)

Tirion, W., *Sky Atlas 2000.0*, 2nd edition. Cambridge University Press (1998)

Webb, T.W., *Celestial Objects for Common Telescopes* (2 volumes). Dover (1962)

## Other resources

American Association of Amateur Astronomers
P.O. Box 7981, Dallas, TX, 75209-0981, USA
www.corvus.com
The AAAA website includes an online newsletter, astronomical guidebook, planetary data, images, and links to other sites.

*The Astronomer*
www.theastronomer.org
From its 1960s origins as a newsletter for observers, TA, as it is known, has grown into an important medium for amateur–professional collaboration, including supernova search programs.

The Astronomical League
11305 King Street, Overland Park, KS, 66210-3421, USA
www.astroleague.org
The Astronomical League is composed of more than 240 local organizations

from across the United States. Their site has links to lists of objects, including challenges such as the Herschel 400.

*Astronomy*

Kalmbach Publishing Co., 21027 Crossroads Circle, P.O. Box 1612, Waukesha, WI, 53187-1612, USA

www.astronomy.com

Popular magazine aimed at both amateurs and professionals. It regularly carries articles on deep sky observing.

*Astronomy Now*

Pole Star Publications Ltd., P.O. Box 175, Tonbridge, Kent, TN10 4ZY, UK

www.astronomynow.com

The leading astronomy magazine in the UK. It regularly carries articles on deep sky observing.

British Astronomical Association

Burlington House, Piccadilly, London, W1J 0DU, UK

www.britastro.org/main

The BAA is the premier observing body for amateur astronomers in the UK and has several specialist sections, including one devoted to the deep sky.

Deep Sky

www.deep-sky.org

This site provides a user friendly, graphically based way of exploring the deep sky. It is solely dedicated to the science and enjoyment of visual deep sky astronomy.

The Interactive NGC Catalog Online

www.seds.org/~spider/ngc/ngc.html

This site has listings for all the New General Catalogue objects and includes images of many of them.

International Dark-Sky Association

225 North First Avenue, Tucson, AZ, 85719, USA

www.darksky.org

A global organization that campaigns against light pollution.

Megastar

P.O. Box 35025, Richmond, VI, 23235, USA

www.willbell.com/software/megastar/index.htm

High-powered charting software from Willmann-Bell that is well respected by deep sky observers.

The NGC/IC Project

www.ngcic.com/default.htm

A comprehensive site with listings for objects in the New General Catalogue and Index Catalogues.

Redshift

www.redshift.de

A popular desktop planetarium software program containing a great deal of useful information.

Royal Astronomical Society of Canada

136 Dupont Street, Toronto, ON, M5R 1V2, Canada

www.rasc.ca

The principal astronomical society in Canada. Their annual *Observer's Handbook* contains useful listings of objects.

SEDS Messier Database

www.seds.org/messier

Extremely useful site listing details of all the Messier objects, with links to additional deep sky information and catalogues.

*Sky & Telescope*

Sky Publishing Corporation, 49 Bay State Road, Cambridge, MA, 02138-1200, USA

www.skyandtelescope.com

The oldest established and still one of the most popular astronomy magazines. Their site contains many resources, including information for the deep sky observer.

Space Telescope Science Institute

www.stsci.edu

Their site provides a range of resources, including catalogues and surveys, online publications, software and a picture gallery.

Starry Night

Imaginova Canada Ltd., 284 Richmond Street East, Toronto, ON, M5A 1P4, Canada

www.starrynight.com

Star-charting software available at several levels of complexity.

TheSky

Software Bisque, 912 12th Street Golden, CO, 80401-1114, USA

www.bisque.com

Popular desktop planetarium software that is available at differing levels of complexity.

The Webb Society

www.webbsociety.freeserve.co.uk

International organization that encourages deep sky enthusiasts to test the limits of the their equipment and observing ability. They publish useful guides, reports, and articles.

# INDEX

## OBJECT INDEX

## INDEX OF M NUMBERS